A geomorphological reconnaissance of
Sumatra and adjacent islands (Indonesia)

Verhandelingen of the Royal Dutch Geographical Society (K.N.A.G.) 1

A geomorphological reconnaissance of Sumatra and adjacent islands (Indonesia)

H. Th. Verstappen

International Institute for Aerial Survey
and Earth Sciences (I.T.C.) – Enschede

Wolters-Noordhoff Publishing Groningen 1973

Publication of this volume has been made possible by a grant from The Netherlands Organization for the Advancement of Pure Research (Z.W.O.)

On the cover: fragment from the map 'Indiae Orientalis Nova Descriptio', first published in: Mercator - Hondius - Janssonius, *Atlantis Maioris Appendix,* Amsterdam 1630.

© 1973 Wolters-Noordhoff nv Groningen, The Netherlands.

Niets uit deze uitgave mag worden verveelvoudigd en/of openbaar gemaakt door middel van druk, fotokopie, microfilm of op welke andere wijze ook, zonder voorafgaande schriftelijke toestemming van de uitgever.

No part of this book may be reproduced in any form, by print, photoprint, microfilm or any other means without written permission from the publisher.

ISBN 90 01 90938 8

Contents

PREFACE
I INTRODUCTION
   a. General character 1
   b. Sources of information 1
II MAIN GEOMORPHOLOGICAL CHARACTERISTICS OF SUMATRA
   a. Structure 6
   b. Climate 9
   c. Vegetation 12
   d. Weathering 13
III THE GEOMORPHOLOGICAL MAP AND DESCRIPTION
   a. The map 14
   b. The description 16
IV THE ISLANDS OF THE SUNDA SHELF, EAST OF SOUTHERN SUMATRA
   a. Peneplain 18
   b. Residual hills 19
   c. Quaternary features 20
V SOUTHERN SUMATRA
   a. Introduction 23
   b. The Bengkulu Block 24
      1. General features 24
      2. Drainage pattern 25
      3. Raised beaches 27
      4. Volcanoes 29
   c. The Median Graben 31
      1. General features 31
      2. Semangka Section 31
      3. Ranau Section 33
      4. Mekakau-Tandjungsakti Section 36
      5. Keruh-Musi Section 37
      6. Ketaun Section 37
   d. The mountainous areas to the East of the Median Graben 39
      1. General features 39
      2. Semangka and Ratai Blocks 40
      3. Huluwaisamang granites and surrounding volcanoes 42
      4. Gedongsurian graben and adjacent block mountains 42

        5. Komering Gap and Garba Mountains   43
        6. Pasemah and Gumai Mountains   44
        7. Kaba volcano   46
    e. The low areas of eastern Sumatra   48
        1. Sekampong Block   48
        2. Sukadana Basalt plateau   50
        3. The narrow alluvial plain in the South   51
        4. Palembang Lowlands, general features   51
        5. Drainage pattern of the Palembang Lowlands   52
        6. Ancient coastlines   55
        7. The northwestern parts of the peneplain   56
VI CENTRAL SUMATRA
    a. Introduction   58
    b. The Block Mountains to the West of the Median Graben   60
        1. The continuation of the Bengkulu Block   60
        2. Indrapura plain   60
        3. Padang Section   61
        4. Fluvio-volcanic fans of the Padang plain   62
        5. Lake Manindjau cauldron   63
        6. Ophir Section and adjacent lowlands   64
    c. The Median Graben   66
        1. The parts southeast of Lake Kerintji   66
        2. Kerintji intramontane plain   67
        3. Batang Hari Section   70
        4. The Baruh and Atas lakes   71
        5. Talang volcano   74
        6. Solok-Singkarak Section   75
        7. Bukittinggi ignimbrite plateau   78
        8. Masang-Sumpur Section   81
    d. The Mountainous Areas to the East of the Median Graben   82
        1. The southern parts   82
        2. Kerintji and Tudjuh volcanoes and related fluvio-volcanic features   83
        3. Mountains between the Kerintji and Marapi volcanoes   85
        4. Marapi and Malintang volcanoes   87
        5. The northernmost parts   89
    e. The low areas of eastern Sumatra   89
        1. Peneplain   89
        2. Drainage pattern South of the Kwantan River   90
        3. Drainage pattern North of the Kwantan River   91
        4. Recent tectonics in the lowlands   91
        5. Beach ridges   92

VII NORTHERN SUMATRA
- a. Introduction   93
- b. The Block Mountains to the West of the Median Graben   94
  1. Malintang volcanic complex   94
  2. Sorikmerapi volcano and adjoining faults   94
  3. The Natal plain and its hinterland   99
  4. Strongly block-faulted areas between Natal and Tapanuli Bay   100
  5. Hinterland of Tapanuli Bay and areas West of Lake Toba   100
  6. Singkel-Meulaboh Section   101
  7. Northernmost parts of the western block mountains   102
- c. The Median Graben   103
  1. The transition zone toward central Sumatra   103
  2. Deposits and terraces in the graben   106
  3. Padangsidempuan Section   106
  4. The graben South of Lake Toba   108
  5. The Median Graben West of Lake Toba   110
  6. Kutatjane (Alas) plain   111
  7. Blangkedjeren basin   116
  8. Deposits and terraces in the Blangkedjeren basin   117
  9. Northernmost parts of the graben   118
- d. The mountainous areas to the East of the Median Graben   119
  1. Sumper graben   119
  2. Asik – UluAer fault zone   119
  3. Southern part of the Toba plateau   122
  4. Lake Toba graben   124
  5. The sliver along the western side of the Toba graben   126
  6. Diatom deposits of Samosir   126
  7. Young faults and grabens on Samosir   128
  8. Fault-scarp morphology of the east coast of Samosir   129
  9. Fault-scarp morphology of the Sibolangit peninsula   132
  10. Lake terraces on the Sibolangit peninsula and along the southern lake shore   135
  11. Asahan valley   140
  12. Karo Highlands and surroundings   146
  13. Stratovolcanoes bordering on the Karo Highlands   148
  14. Mountains between the Karo Highlands and the Gajo Districts   149
  15. Gajo Districts   150
  16. Laut Tawar and the Peusangan fluvio-volcanic terraces   151
  17. Northernmost areas and the island of Weh   153
- e. The low areas of eastern Sumatra   154
  1. Padang Lawas and the plain E of the Toba ignimbrites   154
  2. Alluvial plains in the North and their hinterland   154

VIII THE ISLAND FESTOON TO THE WEST OF SUMATRA
- a. Introduction   160
- b. Outline of the geological development   161
- c. Enggano   162
- d. The Mentawei Islands   164
    1. General features   164
    2. South and North Pagai and Sipora   165
    3. Siberut   166
    4. Submergent coasts and islands to the East   167
- e. The northern part of the Island festoon   170
    1. Batu Archipelago   170
    2. Nias   171
    3. Banjak Islands and Simalur   173

IX BIBLIOGRAPHY

Preface

The request to write a preface for Dr. Verstappen's monograph, A Geomorphological Reconnaissance of Sumatra and Adjacent Islands, has various attractive aspects. Firstly, as former president (1968–1971) of the Koninklijk Nederlands Aardrijkskundig Genootschap, in short the Society, I welcome this study as the first volume of a new Publications Series of the Society, the so-called Verhandelingen. Publication of this new series was strongly recommended by the committee of wise men that prepared the merger between the three Geographical Societies in the Netherlands, which was effected in 1967. The merger resulted in a New Style Society and reduced the number of Dutch geographical journals from three to two, both of them of a predominantly social and economic geographical nature, thus limiting the publication space for physical and historical geographical articles. The recommendation of the merging committee to start a new series of publications aimed at restoring the balance and re-establishing publication possibilities for lengthy articles and studies which could not be accommodated in the two remaining journals. What the right hand took away, the left hand generously, though vaguely, returned.

It was the task of the first administration of the New Style Society to materialize the recommendation of the merging committee. This realization took us a great deal longer than was anticipated, due not so much to the lack of manuscripts as to the time consuming procedure of overcoming financial problems to get the series started. The manuscript of the present book, generously illustrated with photographs, field sketches, maps and diagrams, was presented to the Society in 1968. As its own funds were inadequate to publish the Verhandelingen on a private basis, and as apparently no publishing company was willing to undertake the risk, the Society had to negotiate for financial support, which eventually was granted by the Nederlandse Organisatie voor Zuiver Wetenschappelijk Onderzoek (ZWO). Prior to and during these negotiations, the manuscript was critically read and approved for publication by an ad hoc editorial committee. Serving on this committee, on the invitation of the Society, were Professor Dr. J. J. Dozy, Department of Mining, Technical University, Delft; Professor Dr. A. J. Pannekoek, Geological and Minerological Institute of the State University in Leiden; and Dr. A. L. Simons, Laboratory of Physical Geography and Soil Science of the University of Amsterdam, all with long experience in Indonesia. The Society is greatly indebted to these experts who gave freely of their time and made valuable suggestions for improvement of the manuscript.

As former Head of the Geographical Institute of the Topographical Survey in Indonesia I gladly welcome the present study as the crowning touch of Verstappen's exploration as staff member of the Institute from 1949–1957. Due to the generous attitude of the Director of the Topographical Survey, the present Brig. Gen. R. M. Soerjosoemarno, Retired, Djakarta, the climate for exploration for Dutch geographers in the Survey was extremely favorable in the years directly following the transfer of Sovereignty (1949). Young Verstappen, urged by a fervour of exploration and prepared to take the risks and hazards of travel in turbulent post-war Indonesia, eagerly seized the opportunity. In the period 1949–1957 he undertook an impressive series of geomorphological reconnaissance tours, mostly accompanied by Indonesian students or staff members of the Geographical Institute, throughout Indonesia, particularly on Java, Sumatra, and in the Moluccas. An extensive list of publications beginning with his doctoral thesis on Djakarta Bay, A Geomorphological Study on Shoreline Development up to and including the present volume, bear evidence of his travels. Of his various tours those to Sumatra in the years 1951, 1954 and 1955, each of them lasting 3–4 months, were a memorable achievement. With pleasure I recall the many discussions during the planning phase of these tours on the selection of accompanying staff and equipment, on the travel routes and transport problem (canoe, horseback, jeep) and on the possibility of a military escort for the ardent explorer, which he always refused. Five times in succession we saw him off at the harbour of Tandjung Priuk sailing for Sumatra and hoped he would return safely, which eventually he did.

To the Society that functioned throughout a greater part of its lifetime as the patron society for exploration, promoting some expeditions and lending prestige to others, publishing Verstappen's exploration results is the continuation in a modern form of an old tradition.

Finally, the present study should be welcomed as a valuable addition to the scarce geomorphological literature on Indonesia. It may be true that geologically the island of Sumatra, with the exception of Atjeh is one of the better explored parts of Indonesia, however, from a geomorphological point of view it is as virginal and unexplored as the rest of the country. The wide significance of geomorphological studies and its relations to human affairs, particularly in environmental studies on behalf of development planning are widely recognized today. On Sumatra itself the critical importance of the geomorphological factor has become apparent, e.g. during the resettlement of Javanese farmers before and after, and of army veterans after, World War II in the southern part of the island and, more recently, in the engineering project of the Trans-Sumatra Highway, the location of which was recently planned on the basis of aerial photographs. The hope may be expressed that Dr. Verstappen's study will contribute to a better insight into the environment of human activity of Sumatra. Moreover, it will undoubtedly stimulate further geomorphological exploration to the benefit of future development projects on the island.

F. J. Ormeling

I Introduction

a. General character

The island of Sumatra, sixth in size among the world's islands with an area of 434,000 square kilometres, is the largest of a long island arc, called the circum-Sunda arc. Both the outline and the relief of Sumatra are intimately related to its location in this tectonically active zone. The island is many times longer than wide (1650 km versus a maximum width of 350 km) and the width of its main mountain chain, the Barisan Mountains which extend from one end of the island to the other along a nearly straight line, is nowhere more than about 135 km.

The relief of the mountain range is strongly influenced by fault movements accompanied by volcanism, especially along the Median Graben or Semangko fault zone which extends, like the Barisan range, over the entire length of the island. Important vertical movements have locally raised the mountain chain to present heights of more than 3000 m, notwithstanding the strong weathering and erosion resulting from the tropical climate. Quaternary (even Holocene) tectonic movements and interacting glacio-eustatic sea level changes have left distinct traces.

The structure of the island is distinctly asymmetric. E of the mountain chain extends a sedimentary basin with young folding; this is bordered in its turn by an alluvial plain of recent origin, underlain by the stable Sunda platform consisting of older rocks. This recent accretion (Tjia et al., 1968) has given the E coast an irregular outline contrasting with the almost straight W coast.

b. Sources of information

The island of Sumatra was one of the author's main objects of geomorphological research during his stay in Indonesia from 1949 to 1957; the first visit to the island was made in 1951, when a few weeks were spent in the Padang Highlands. Extensive field work was carried out in 1954 and 1955 on behalf of the Geographical Institute of the Topographical Survey in Djakarta. These studies were meant to serve primarily as support for the pedological reconnaissance of the island, and furthermore aimed at an improved representa-

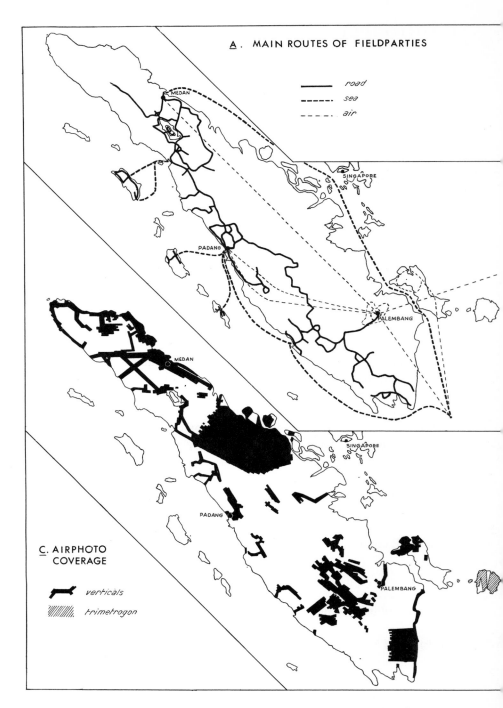

Fig. 1. Sources of geomorphological information on Sumatra: A. The author's main field routes; B. Coverage of topographical maps; C. Coverage of aerial photographs (1957); D. Coverage of geological and soil surveys.

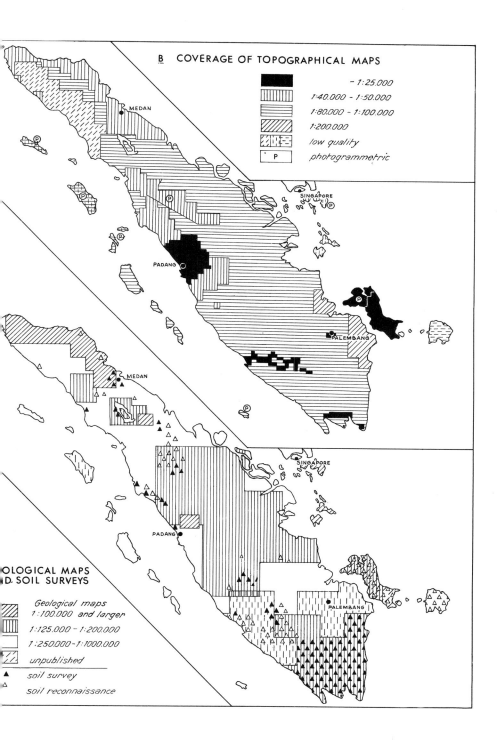

tion of the land forms on small-scale maps. The main routes taken by the field parties are indicated in Fig. 1A. The work was concentrated on the Barisan Mountains in the W; the rather monotonous eastern lowlands were visited only briefly.

Southern Sumatra was reconnoitred in March and April of 1954. Part of this investigation was done at the request of the Archaeological Service and concerned the physical setting of the ancient Srivijaya kingdom in Palembang. Another part of the work was done in connection with the rapidly expanding pioneer settlements in the area, and this was the author's first experience in applying geomorphology to development projects (Verstappen, 1956). Central Sumatra and in particular the Kerintji intramontane graben was studied in August and September of the same year. A short communication concerning the results has appeared elsewhere (Verstappen, 1955). The islands to the W of Sumatra were the subject of field work carried out from March to May of 1955. In selecting the areas to be studied, preference was given to parts not visited by earlier investigators. Finally, field work was carried out in northern Sumatra from August to November, 1955. Considerable time was spent in the Lake Toba area, and a short paper on this and some other 'volcano-tectonic' depressions has already been published (Verstappen, 1961). Unfortunately, for reasons of security the author was not allowed to visit the northernmost tip of the island (Atjeh). The field investigations ended in the Blangkedjeren intramontane basin. Some visual aerial observations could be made during regular and special flights over several parts of the island, and these provided supplementary geomorphological data.

It is evident that, notwithstanding the elaborate field work, the information gathered on this vast island was insufficient to allow a geomorphological description of any completeness. Other sources of information were therefore drawn on and have contributed (appreciably) to this description of the island and the accompanying geomorphological map.

Data collected in more detailed studies carried out by the author in a small number of characteristic areas of limited extent have been added to this framework. The (semi-)detailed geomorphological maps prepared for these areas are included in this paper.

All aerial photographs of the island existing at the time of the study were studied stereoscopically, and yielded a large amount of geomorphological data for the areas concerned. Unfortunately, the airphoto coverage of the island was and still is very incomplete and rather haphazard. Narrow strips cover some major valleys and intramontane plains in the Barisan Mountains, the more continuous coverage pertaining to the oil basins of the geomorphologically less interesting eastern hills and lowlands (fig. 1 B).

Analysis of all sheets of the topographical map series covering the island (Fig. 1 C) was of the utmost importance for the completion of the geomorphological map. The western and central parts of northern Sumatra have not yet been surveyed; only sketchy and incomplete 1 : 200,000 maps exist, and consequently the quality of the geomorphological map of these areas is poor.

Geomorphological literature concerning the island is almost nonexistent, but some information can be obtained from geological studies and soil surveys, the coverage of which is indicated in Fig. 1D. Descriptions and reports given by volcanologists, topographers, and other explorers have also been useful. The outstanding geological description of the island given by Van Bemmelen in his book 'Geology of Indonesia' (1949) deserves special mention.

The maps 1a–d and the list of references give full information on the sources used and thus are essential for evaluating the reliability of the geomorphological map for all parts of the island.

It goes without saying that this study should be considered as a first attempt to unravel the geomorphological outlines of the island. It will have served its purpose if it proves to be of some use as a framework for future, more detailed studies, and if it can serve as an aid to soil scientists and other specialists engaged in the survey of natural resources and planning.

Until now, Indonesia has had a small-scale geomorphological map only for the nearby island of Java. This map was published on a scale of 1 : 1,000,000 by Pannekoek, and accompanies his well-known paper *Outline of the geomorphology of Java* (1946). It is hoped that in the future, similar geomorphological maps will also become available for other sections of the partly sub-marine Sunda Mountain system, which stretches from Burma in the NW to the easternmost parts of Indonesia.

The author wishes to thank Brig. Gen. R. M. Soerjosoemarno, then Director of the Topographical Survey of Indonesia, for his never-failing support of the field work related to this study and Col. A. Hamid and his staff of the Geographical Institute, for valuable assistance both in the field and in the Institute. He is also greatly indebted to Prof. F. J. Ormeling, at the time Head of the Geographical Institute, for his assistance in launching the survey and for his encouragement during its execution, and to Prof. J. J. Dozy and Dr. A. L. Simons for their critical reading of the manuscript. Special thanks are extended to Prof. A. J. Pannekoek, who devoted so much time to preparing the additional maps in Figs. 10, 12, 14, 17, 18, 24, 38, 47, and whose extensive knowledge of Indonesian geomorphology supplied invaluable information.

II Main Geomorphological characteristics of Sumatra

The land forms of Sumatra are the product of the continuous interaction of structural and climatological factors. The effects of important crustal movements of the recent past and of the present and the related volcanic activity are counterbalanced by the rapid degradation and aggradation for which the humid tropical climate is responsible.

a. Structure

The structural factors (Van Bemmelen, 1949) are determined by the fact that Sumatra forms part of the Sunda Mountain system, which extends from Burma to the South Moluccas. This explains the length and comparative narrowness of the island. The volcanic inner arc of this mountain system is represented in the Sumatra section by the Barisan Range, which extends in a NW–SE direction at a short distance from the W coast of the island. This geanticlinal upwarp forms the backbone of Sumatra. It is separated by an interdeep (locally over 2,000 m deep) from the partly submerged non-volcanic outer arc, which is represented by the island festoon to the W of Sumatra and comprises Enggano, the Mentawei Archipelago, Nias, and Simalur. The islands of the non-volcanic arc are connected to Sumatra by three submarine swells dividing the interdeep into separate basins. The southernmost of these connections occurs between Enggano and South Pagai, near the islet of Mega. A second is to be found E of the Batu islands, the island of Pini marking its highest part. The northernmost is situated between Nias and Simalur and is marked by the Banjak islands. A deep sea trough was formed at the ocean side of this non-volcanic arc.

A geosynclinal hinterdeep occurs on the northeastern side of the Barisan range, where the low hills and the alluvial plains of Sumatra are located. The alluvial plains along the eastern coast become gradually narrower toward the N. Sedimentation was evidently less intense here, or at least could not keep pace with the subsidence. The northernmost parts of the geosynclinal zone are largely submerged, and as a result the Malacca Straits become broader at the expense of the width of the island of Sumatra.

The southeasternmost part of Sumatra, together with the islands of Bangka, Belitung, and the Riau/Lingga group, forms part of the more or less stable Sunda shelf and thus differs completely from the orogenic and unstable remainder of the island. Some small granitic outcrops, e.g. near Palembang, and also evidence provided by borings in the northeastern Lampong Districts, confirm that the Sunda massif occurs at a rather shallow depth in southeastern Sumatra.

Photo 1, showing a relief prepared by the Geographical Institute in Djakarta, clearly reveals these main geomorphological features of Sumatra. The island is two or three times broader in the S than in the N, due to the increasing depth of the East-Sumatran geosyncline toward the NW and the presence of the Sunda massif at a shallow depth in the southeastern tip of the island.

The highly asymmetric cross-section of the island, already referred to, is also a striking feature. The southwestern side is generally steep, and only insignificant alluvial plains occur. The coastline is rather straight and often rocky. The northeastern side of the island, on the contrary, is predominantly

Photo 1. Relief model of Sumatra, prepared at the Geographical Institute, Djakarta.

a lowland. The coast here is low and has many indentations. This asymmetry is evidently attributable to the development of the hinterdeep geosyncline to the E of the geanticlinal Barisan range. A factor of secondary importance is a certain preference shown by the pre-Tertiary rocks and granitic intrusions for the eastern side of the geanticlinal crest, especially in southern Sumatra.

The geology of the island has been extensively dealt with by Van Bemmelen (1949), and a brief summary will therefore suffice here. Reference to the geological literature on specific parts of the islands will be made where the relevant material is dealt with.

The Barisan range consists largely of Upper Palaeozoic and Mesozoic sedimentary rocks. Isoclinal folding and overthrusting toward the SW has been observed (Tobler, 1919; Kugler, 1922) but the existence of larger nappes has been questioned by Katili (1970b). Intrusions of granite and granodiorite occur. The uplift of the range started in the Lower Palaeogene, and the beginning of the faulting and rift-forming activity dates back to this period. Subsidence occurred during Oligo-Miocene times and the Barisan range became partly submerged again. Sedimentary rocks of this age now occur to the W and E of the range and locally also in its interior. At that time, andesitic volcanism was important in southern Sumatra. An important geanticlinal upwarp occurred in the range during the Middle Miocene. Block faulting and intense, acid volcanic activity accompanied this renewed orogenic activity. Granodioritic intrusions are also known from this period. No important subsidence of the range has since occurred, but prolonged base-levelling must have taken place prior to the latest block-faulting movements and consequently the surface of these blocks is rather flat, at least where they are not too strongly dissected. A final period of orogenic activity is reported to have taken place in the Plio-Pleistocene: block-faulting and rift formation in the Barisan continued with renewed force. The Semangko fault zone, a broken-up longitudinal (median) graben located along the crest of the geanticline, is a remarkable feature. Horizontal movements also seem to have occurred here. According to Katili (1970a,b) it is a dextral transcurrent fault zone; Tjia (1970a) mentions both dextral and sinistral offsets.

The block mountains form the foundations of the Quaternary volcanoes, which are predominantly associated with faults. The volcanic products cover large areas of this 'pre-volcanic surface' in the High Barisan, particularly in the S. Simultaneously with the block-faulting of the Barisan there was important sedimentation and subsequent folding in the geosyncline of eastern Sumatra. The general trend of the folds is parallel to the axis of the island. Continuous aggradation in the lower parts accompanied by degradation in the higher parts, resulted in the present low rolling hills of this zone. A broad alluvial plain originated in the E. The present outline and relief of the island thus came into being, but important tectonic movements continue up to the present day.

The islands of the non-volcanic arc situated to the W of Sumatra consist largely of thousands of metres of Tertiary deposits, the bulk of which are not very resistant to erosion. Outcrops of pre-Tertiary rocks are rare; ultrabasic rocks occur locally. The uplift of this zone, due to the isostatic rise of its root, started rather recently, in the Quaternary.

The islands to the E of southern Sumatra, forming part of the Sunda shelf, are mainly composed of Mesozoic and Upper Palaeozoic sediments. Granitic intrusions are important features. The whole area has been largely base-levelled, but the arrangement of the islands clearly indicates the structural trendlines – generally striking NW–SE – of this old land mass.

b. Climate

The climatological factors prevailing in the humid tropics lead to rapid degradation of the land forms, especially if rather unresistant and relatively young sedimentary rocks and volcanic products are mainly concerned, as is the case on Sumatra and adjacent islands. Planation therefore proceeds much more rapidly than in a moderate climate, and as a consequence the land forms often rather clearly reflect the more recent crustal movements. Older tectonic movements are, of course, less easily traced geomorphologically, because of the rapid base-levelling.

It is relevant to devote some attention to the climatological conditions of Sumatra and their impact on vegetation, since such considerations contribute considerably to an understanding of the development of land forms. Most parts of Sumatra are characterized by abundant rainfall distributed rather regularly over the year. The average annual temperature in the lowlands lies between 26 and 27° C. The climate is of the tropical rainy type (Köppen's A climates). Because there is no pronounced dry season and the temperature in the lowlands and hills during the warmest months remains above 22° C, the *Afa* climate is by far the most common. In the higher parts of the mountains *Af*, *Cfhi*, and *Cfi* climates occur. *Asa* and *Awa* climates are rare, but the *Ama* (and *Am*) climate is the second most important climatic type of the island. This means that the precipitation during one or more months is less than 60 mm, but that rainfall during the other months is so abundant that the vegetation is little affected by the dry period. It is therefore clear that physical weathering, which requires dryness to be effective, is of little importance in Sumatra.

F. H. Schmidt and J. H. A. Ferguson (1952) used the ratio Q between the number of dry months (< 60 mm) and the number of wet months (> 100 mm) as a criterion for their classification of the Indonesian climates. From their data, which are interesting for climatic-geomorphological studies and are given in Fig. 2B, it appears that Q is less than 14.3 per cent in the greater part of the island (Class A); values of up to 33.3 per cent (Class B) occur in

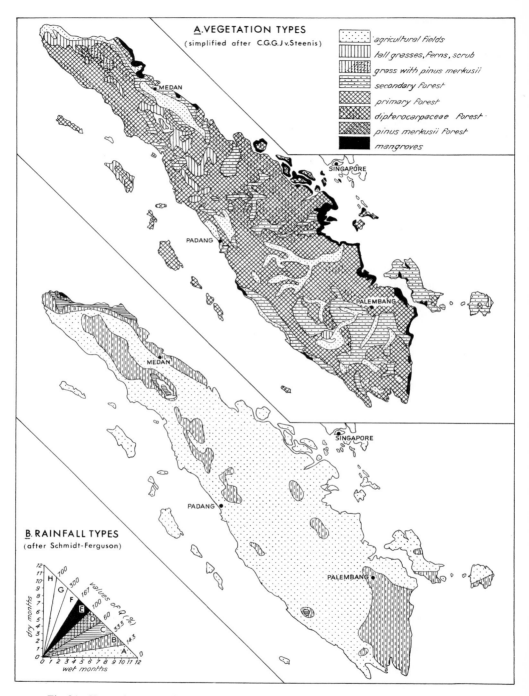

Fig. 2A. Vegetation types of Sumatra, after van Steenis (*Atlas van Tropisch Nederland*, 1938).

Fig. 2B. Rainfall types of Sumatra; based on ratios between wet and dry periods. After Schmidt and Ferguson, 1951.

the SE, in the Barisan Mountains, and along the northeastern coast; still higher ratios are only found in the extreme N.

The driest months are normally associated with the easterly monsoon, but a small area in the northern part of the island has the Asiatic type of rainfall, with February as the driest month. The main rainy season usually falls during the transition period following the easterly monsoon, a rainy period of secondary importance occurring during the preceding transition. In the southeastern part of the island the rains of the transition periods are immediately followed by the monsoonal rains, and thus there is only one relatively dry period, as in nearby Java.

A matter of particular interest is that on Sumatra not only do the climatic conditions have an impact on the geomorphology of the island but conversely the relief also has a strong influence on the climatological phenomena. This of course also holds for other areas, but it is particularly striking on this island because the Barisan Mountains lie approximately perpendicular to the monsoonal winds. This results in wide local variations in annual precipitation, ranging from 1127 mm (Atjeh) to 6134 mm (SW coast). Monthly values of 400 to 600 mm occur along the western coast of central and southern Sumatra during November at the windward side of this mountain barrier, whereas low rainfall figures are recorded at several stations located in the intramontane grabens and basins.

The winds are also strongly influenced by the presence of the Barisan Mountains. The force of the monsoonal winds to the E of the range is generally small, with the exception of the southeastern part of the island. Strong westerly and southwesterly winds occur in the mountains, however, blowing around the volcanic peaks, through the passes, and over the plateaus, thus rendering the climate there unpleasantly cool. These strong winds also occur in the intramontane graben, where the lakes may become very rough. The Depéq winds at Lake Tawar (Atjeh) are notorious in this respect.

More important for our purpose is the descent of warm dry föhn winds to the E of the mountain barrier, particularly where the mountains are low and narrow and where the slope of the eastern plains is rather gradual. The Bohorok winds of Deli are especially ill-famed, because of the damage they do to the tobacco fields. They blow from the W and SW, and occur mostly between May and October. Similar föhn winds also occur in other parts of the island. The worst is undoubtedly the one in the Padang Lawas area, located to the SE of the Toba–Batak plateau in northern Sumatra. This wind blows almost continuously during several months of the year because of the easy passage provided by the low and narrow parts of the Barisan Mountains bordering the area on the W. The soil dries out completely and extensive grass plains occur instead of the forest cover otherwise so characteristic for the Sumatran landscape. Physical weathering unquestionably plays a role in this region.

c. Vegetation

The vegetation pattern resulting from these climatic conditions and also partly from human interference with the natural conditions, particularly in the southeastern parts of the island, deserves attention because the vegetative cover has a strong influence on the conditions of surface run-off.

The vegetation map prepared by van Steenis (Fig. 2A) shows that the island is predominantly covered by primary rain forest. Secondary forests also occur, especially in southern Sumatra, on Bangka, and on Belitung, primarily as the result of shifting cultivation. Grass, scrub, and ferns are reported from parts of Nias, the Toba area, and other areas in northern Sumatra. Agricultural fields are preferably located in river valleys, on intramontane and other plains, in coastal areas, etc., but are rapidly expanding in several other parts of the island. Mangrove vegetation is important along the E coast.

As far as the vertical zoning of the vegetation is concerned, it should be noted that almost the whole of the island lies below the tree line; the highest block of the Barisan Mountains, Mt Löser, located in the N, reaches 3,381 m. The highest volcano of the island (and of Indonesia as a whole), Mt Kerintji (3,805 m), approaches the tree line. Its peak and those of some other active volcanoes are the only parts of the Barisan Mountains virtually uncovered by vegetation, but this is due to the rocky nature of these parts rather than to the climate. It should be noted that the moss forest occuring above about 1,500 m reduces the rainwash on the steep mountain slopes to a minimum, which makes these forests a geomorphological factor of importance. This forest is considerably more effective in this respect than even the primary forest covering the lower hills. Approaching the tree line, the subalpine vegetation becomes lower and less dense and is gradually characterized by Rhododendron, Vaccinium, Gaultheria, Leptodermum, and other genera, until finally we arrive at the open fields or '*Blangs*' well-known from the Löser. This vegetation probably is not a true climatic climax.

The grass and scrub plains occurring at lower altitudes in northern Sumatra are due to poor soil conditions (e.g. the Toba tuffs), the strong and often dry winds mentioned earlier in this paper, and a number of other factors. Interception of the abundant rain is evidently less effective in these areas. It is noteworthy that in many places the vegetation is denser and higher in the valleys. The lower ground-water level in the ridges and the heavier evaporation due to the strong winds in these exposed areas account for this situation.

d. Weathering

Among the exogenous processes playing a leading role in landscape development in Sumatra, chemical weathering must be assigned first place. Its rapid progress is due to the abundant rainfall and the high concentration of various acids (carbonic acid, humic acid, etc.), carried off in solution by the water. Particularly under the dense forest, the great humidity and high temperatures induce 'hothouse' conditions which accelerate the alteration of the rocks. A thick cover of weathering products is thus present almost everywhere. Physical weathering is usually of little importance, as explained above. The rapid alteration and formation of soil thus provoked is equalled by a rapid dissection due to the great density of the drainage pattern caused by the heavy rainfall. The smaller ravines only contain water after the tropical rain showers. The combined influence of intense dissection and a thick soil cover results in a high frequency of slides. Steep valley slopes are nevertheless usually maintained, owing to the rapid vertical erosion and the conservational effects of tree roots.

Differences in relief due to lithology are in most cases less pronounced than in other climatic zones. An exception is formed by the limestones, which often show the typical 'conical' karst forms well known from many humid tropical regions. As a result of this rapid degradation, uplifted old surfaces are comparatively soon transformed into narrow-crested ridges and small hills, and finally disappear. As a result of the same processes, however, the formation of peneplains proceeds rapidly.

The rapid degradation causes a high silt content of the river water and the rapid growth of alluvial plains. It should be noted, however, that most of the material is weathered so intensely that mainly clayey particles remain. Scree fans and coarse alluvial fans are thus rare phenomena occurring only where coarser-grained material is available, e.g. from volcanic sources. Normally, steep and often slightly concave valley-sides are bordered by plains almost without transition.

The coastal features are influenced by coral growth in clear waters and by mangrove vegetation, particularly along the low eastern coast of the island. Tidal currents also play an important role, as will be discussed below.

III The geomorphological map and description

a. The map

The geomorphological map on a scale of 1:2,500,000 accompanying this paper is an attempt at a small-scale cartographic representation of the main characteristics of land forms according to the modern analytical principles stipulated for geomorphological maps. The map was drawn by J. Bloemendaal, cartographer at the International Institute for Aerial Survey and Earth Sciences (I.T.C.), Enschede. It is evident that for this type of map not only a high degree of generalization is unavoidable, but the nature of the information included will also be somewhat different from that indicated on detailed maps. More stress is naturally put on the main structural features, whereas the detailed relief forms and the processes underlying them, which are of prime importance on detailed geomorphological maps, cannot be adequately indicated. Notwithstanding the obvious limitations, maps of this type have already been constructed in several countries, with the use of various legends. The value of these modern small-scale geomorphological maps lies above all in the wealth of comprehensive information they provide. They have replaced the incomplete sketchy maps and the pictorial physiographic maps produced earlier, and will contribute to geomorphological reconnaissance studies in much the same way as modern detailed geomorphological mapping techniques have improved detailed surveys.

The 1:1,000,000 International Map of the World has been used as a topographical basis for the map, to which the 200 m depth contour was added from charts. Not all geographical names mentioned in this paper could be included in the map, of course, but map coordinates of many features are given to facilitate their location; the coordinate numbers are indicated between brackets. Horizontal (X-) coordinates are given first. The numbers correspond with the grid numbers in the margin of the 1:2,500,000 map. Where necessary, the reader is advised to consult, for instance, the 'Atlas van Tropisch Nederland' or the relevant sheets of the International Map of the World 1:1,000,000.

The geomorphological information derives from the sources mentioned on p 1. Morphometric and morphographic as well as morphogenetic and morphochronologic data are included, and some lithological information has also been added. It need hardly be said that such diversified information is not yet available for many parts of the island; a system was therefore devised that made it possible simply to refrain from indicating the above-mentioned aspects where information was lacking. Since some of the existing geological and other maps are rather inaccurate, the problem of whether or not the lithological and other information should be incorporated into the geomorphological map often arose. Therefore, the amount of lithological information was purposely limited, which is not too serious a handicap because the effect of lithology on the land forms is considerably weaker in the humid tropics than in other climatic zones.

Morphometric data on the map include heights of the most important mountains and barbed lines indicating two classes of valley depth. To arrive at a satisfactory representation of the morphographic aspects, the map symbols were selected so as to resemble the actual relief forms as well as possible. Special attention was paid to morphogenetic information; colours are reserved for this purpose. The legend shows that three main morphogenetic groups are distinguished, *viz.*: forms of exogenous origin, forms of structural origin, and forms of volcanic origin. The distinction is not always easy to convey, however, and a certain amount of overlapping occurs in many cases. For example: block mountains can be formed in a pre-existing peneplain and both exogenous and structural origin should then be mapped. Transitions between structural and volcanic forms occur, e.g. where tuff deposits partly bury the block mountains or when block mountains are formed in volcanic terrain. These problems are very numerous in Sumatra, because the island is characterized by intense young tectonic movements, for which special symbols were added to the map, and by active volcanism.

The larger stratovolcanoes are distinguished according to their degree of dissection. The youngest volcanoes are little dissected; the older, extinct cones are less well preserved, and the intense dissection of their slopes has resulted in the formation of distinct but irregularly outlined radial ridges. These characteristics are depicted in the geomorphological map by the use of straight radial lines for the young cones and slightly undulating radial lines for older ones. The caldera rims and main ridges of the latter features could often be indicated, even on this small scale. Fluvio-volcanic fans and the vast tuff/ignimbrite plateaus are other features of importance. Various lava forms are also included.

Distinction is made between regions of aggradation and those of degradation. A rough indication of the lithology of the former is made by applying dots for sandy and other coarse-grained deposits and by omitting them for fine-textured materials. Beach ridges, marking former positions of the coast line, have been mapped in this way wherever their existence could be ascertained. The main lithological distinction in the degradational areas is between several types of crystalline rocks. In sediments, limestones are specially indicated because of their characteristic humid-tropical karst forms. Resistant rocks in particular give rise to the 'residual hills' of the monadnock type or to the 'isolated higher ridges' shown on the geomorphological map. These ridges usually form in quartzites or sandstones. Ridges rising above the average level of the peneplain of eastern Sumatra and composed of folded, mostly Tertiary sediments belonging to the East-Sumatran geosyncline, could therefore be mapped. The morphochronology has not been stressed to any great extent, but is indicated by abbreviations (small capitals) where the age of the forms could be established.

Several areas of specific interest are depicted here on larger-scale geomorphological maps, which vary in accuracy. The coloured map of the Talang volcano und surroundings in central Sumatra, shown in Fig. 25, is an example of a modern detailed geomorphological map, based on fieldwork and interpretation of aerial photographs. The map of the fluvio-volcanic terraces near Lake Tawar, in northern Sumatra (Fig. 62), belongs to the same class. Since no field check could be carried out in this area, however, the latter is only a preliminary map, based entirely on photo interpretation, and is therefore printed in only two colours. These two maps were prepared by H. P. Kruyne, cartographer at I.T.C., Enschede. The topographical basis of these two maps was obtained photogrammetrically at I.T.C. with a Wild A-8. The other geomorphological maps included in this publication are less accurate and are accordingly printed in black.

b. The description

The geomorphological description is subdivided regionally into five main parts.

The first comprises the islands to the E of southern Sumatra, forming part of the Sunda shelf. The main features treated successively are the peneplain, the residual hills rising above the peneplain (including similar hills in the alluvial plain of Sumatra), and the Quaternary features.

The description of Sumatra proper is, for the sake of convenience, divided into three main parts, covering southern Sumatra, central Sumatra, and northern Sumatra, which do not coincide with administrative boundaries. The division between southern and central Sumatra has been taken approximately at coordinate 50 of the map, that between central and northern Sumatra at coordinate 30.

Each of these main parts has been further subdivided, because the main structural and morphologic units show little change over the whole length of the island. First, the block mountains along the W coast are considered, then the Median Graben occurring along the culminating axis of the Barisan Mountains; next come the mountainous areas E of the Median Graben, and finally the low areas in the E, including the coastal plain. It should be noted that what is actually the southwest coast is customarily referred to as the W coast and the northeast coast as the E coast.

The fifth and last main part deals with the islands W of Sumatra, which are situated on a submarine ridge parallel to Sumatra and have many features in common.

IV The Islands of the Sunda Shelf, East of Southern Sumatra

Three main land-form types can be distinguished in this area, which comprises from SE to NW the islands of Belitung and Bangka, and the Lingga and Riau archipelagos. The distribution of these land-forms types, i.e. the peneplain, the residual hills, and the alluvial plains, is indicated on the geomorphological map.

a. Peneplain

The peneplain covers by far the largest part of these islands, and at present generally lies 30 to 50 m above sea level, showing a typical senile landscape of gentle hills and ridges with ill-defined divides and broad open valleys filled with Quaternary sediments. The influence of the geological structure on the drainage of Bangka and Belitung is slight except in some localities such as of the Maras Mountains, where the effect of the NW–SE–striking sediments is evident. More pronounced is the influence of the geological structure in the Lingga and Riau Archipelagos, where the peneplain is lower, so that numerous islands occur aligned along faults, fractures, and more resistant beds, and separated by narrow NW–SE–stretching straits scoured by marine currents where fracture zones and less resistant beds are located (Tjia, 1970d).

After its formation, most of the peneplain was submerged and the present sea bottom is thus a product of degradation and aggradation. The occurrence of extensive Quaternary deposits on Bangka and Belitung, first mentioned by Verbeek (1897), and their equivalent, the 'Older Alluvium' in Malaya, which shows a mixture of the characteristics of shallow-water marine deposits and river deposits and was probably laid down in a wide tidal estuary, indicate that the sea once covered even larger areas than it does at present.

Molengraaff and Weber (1919) proved convincingly that the peneplain of the Sunda shelf ran dry during the Pleistocene as a result of glacio-eustacism, and they pointed out that at that time a coherent drainage system existed in the present Java and South China Seas, which can still be traced on the present sea bottom. This brief summary of the development of the peneplain accounts for its characteristic geomorphology. The broad valleys were aggraded as a result of the Post-Pleistocene rise in sea-level. The aggradation is negligible, however, at 40 or more metres above sea-level.

It is known from the investigation of cassiterite placers on Belitung by Van Overeem (1960) that the bedrock topography underneath the alluvial overburden indicates distinct incision of valleys, presumably due to the low sea-levels in the Pleistocene. The rejuvenation is most pronounced in the lower reaches. The extension of these buried Pleistocene valleys can be traced offshore, where placer tin is also mined. Van Overeem furthermore states that the longitudinal profiles of these valleys are not graded but show knickpoints where more resistant beds are traversed. This characteristic is also known from rivers of several other base-levelled, senile areas in the humid tropics. The hill forms are very subdued, in contrast to the bedrock topography of the valleys. Creep phenomena played an important role in the development of slopes; intense chemical weathering disintegrated the rocks, and differential decomposition of shales and sandstones has led to 'collapse' phenomena of the disintegrated sandstones on gentle slopes. The studies of Adam (1932/1933) on the genesis of the cassiterite placers of Belitung and the above-mentioned publication of Van Overeem, particularly the unravelling of geomorphological processes (creep, etc.) have facilitated the exploration for tin on this island.

b. Residual hills

Numerous residual hills rise above the peneplain just described. Photo 2 (scale 1 : 35,000) shows some examples from central Belitung, formed in Permo-Carboniferous sediments. The strike direction is clearly visible. The hill in the extreme NE is situated near the contact with a muscovite-biotite granite. The granites of the island of Bangka usually reach a greater height than the sedimentary rock, but there are exceptions, e.g. the granites near the Kapo River. Mention should also be made, however, of the sandstone monadnock of the Maras Mountains (700 m) in the N and a narrow ridge in the northeastern part of the island near Cape Tuwing. The residual hills of Belitung are partially formed in granodiorites and partially consist of Permo-Carboniferous sediments. The extensive granitic area near Tandjungpandan is, apart from a few minor hills, a low base-levelled terrain.

Information about the residual hills of the Lingga and Riau archipelagos is scanty. The granites of Kundur Island form rather low areas. The residual hill of Papan Island is situated in sedimentaries, and of those on Karimon only two are formed in granite, but Mt Landjut (475 m) on Singkep is granitic. Thus, the various granites occurring on the islands seem to differ greatly in resistance to tropical weathering.

Faber (1954 a, b), who carried out a soil reconnaissance of Bangka and Belitung, also mentions the occurrence of residual hills, and states that those formed in granites show exfoliation features and occasionally have barren and somewhat convex slopes, as for example S of Pangkalpinang. Several granitic hills, according to Faber, are crowned by large rounded granitic blocks or 'tors'. He thus considers them to be relics of former inselbergs. It should be stressed here that the hills often rise from base-levelled areas having the same lithology, as is evident from the hills in Photo 2. In some instances, however, the lithology of the surroundings is divergent. The residual hills thus appear to comprise a/ monadnocks, b/ residual hills on divides ('*Restlinge*'), and possibly c/ inselbergs.

Photo 2. Vertical aerial photograph of central Belitung (scale about 1 : 35,000). The peneplain has an altitude of 20 to 30 m and formed in Permo-Carboniferous sediments except for the extreme NW corner, where granites crop out. The residual hills rising from the peneplain are situated in the more resistant sediments, whose strikes are clearly visible. The higher grounds in the NW are associated with the metamorphic zone surrounding the granite, which is itself non-resistant and formed the source of the tin placers. Note former tin dredgings in the small valley in the NE corner.

c. Quaternary features

A geomorphological element not indicated on the geomorphological map is the Quaternary cover of mainly fine sands. This omission is partly due to the map scale but also to lack of information about its extent. According to

Verbeek, large parts of the islands are covered by it, but Faber mentions its partial disappearance due to strong erosion, leading to the outcropping of weathered rocks. These marine deposits demonstrate the submergence of parts of the island during the Quaternary. The terraces they show are therefore of great interest for the study of Quaternary oscillations of the sea-level, particularly because they occur in one of the tectonically more stable parts of Indonesia. From comparable deposits of the 'Older Alluvium' of Malaya, aggradation up to 70 m (230 feet) above sea-level and possibly even more is reported (Burton, 1962; Courtier, 1962). Terraces are observed at 100, 80, 50, and 25 feet above sea-level. The higher terraces are believed to date from the Lower Pleistocene, although the evidence is scanty.

Van Overeem informs us of the occurrence of a high terrace, 18 m above mean sea-level, in East Belitung. Other remnants are found some kilometres further to the W and SW, at a height of 25 m above mean sea-level. Its base never lies below the present mean sea-level. The main part of the alluvial plains of eastern Belitung is classified by Van Overeem as low terrace and occurs at 6 m above mean sea-level. It is correlated by him with Daly's subrecent lowering of sea-level. A radiocarbon dating of wood fragments originating from the base of the high terrace indicates an age of more than 46,000 years. The whole high terrace is thought to date from the Middle or Lower Pleistocene (Van Overeem, 1960). It is also noteworthy that an abrasion platform occurs off the E coast and that marine sediments often form a level at the same depth. Humate hardplans are known from 18 m below sea-level off the E coast and between 16 and 24 m below sea-level elsewhere. An extensive abrasion platform is believed to exist at 30 m below mean sea-level. The diagram shown in Fig. 3 indicates the situation. A review of Quaternary shore lines of the entire Sunda land is by Tjia (1970 b).

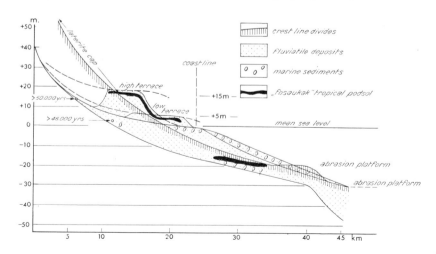

Fig. 3. Profile showing marine terraces above and abrasion platforms below present sea-level in East Belitung. After van Overeem, 1960.

The Recent alluvium surrounding the islands reflects the post-Glacial ingression of the sea. Swamps, in which peat has formed, penetrate the valleys far inland, and mangroves occur at their brackish seaward ends. Sheltered parts of the coast therefore do not have a sharply defined coastline, but more exposed zones are bordered by broad beach ridges stretching from cape to cape. A remarkable feature is the quartz sand area along the eastern coast of Belitung near Manggar. These '*Padangs*' have been described by Faber; they are characterized by poor vegetation.

V Southern Sumatra

a. Introduction

Geomorphologically, the greater part of southern Sumatra can be described as block mountains, many of which are crowned by volcanoes. These block mountains, to the NE of which a folded geosyncline occurs, are generally low in the southeastern part. The surface of the blocks is generally rather flat due to prolonged base-levelling prior to the block-faulting movements. The westward-tilted Bengkulu Block lies W of the Median Graben, and to the E of this rift are the eastward-tilted Lampung (65 x 9), Sekampung (66 x 11), and Sukadana (66 x 12) Blocks (Fig. 14). The highest blocks – and the oldest rocks – occur directly to the E of the rift. The faults separating the blocks all run in a SE–NW direction. Parts of the block mountains do not show a distinct tilt, as for instance in the surroundings of the Gedongsurian Depression (62.5 x 9).

A narrow interrupted alluvial plain separates the Bengkulu Block from the Indian Ocean, whereas along the E coast there is a very broad alluvial plain. It is remarkable that the latter becomes very narrow toward the southernmost tip of the island. This, together with the straightness and divergent direction of the coast here, may indicate the presence of a fault. No folding is found in this southeastern part of the island (Lampung Districts), apparently due to the presence of the Sunda Shelf at shallow depth.

Quaternary volcanoes rise from the block mountains and are predominantly associated with the faults. The volcanic products cover large areas of the High-Barisan, especially to the E of the Median Graben. A number of large stratovolcanoes occur here, forming important elements of the landscape. Higher, older, and somewhat more complex mountains to the E have determined the flow of the fluvio-volcanic products, and so have also had a great influence on the drainage pattern. The eastern blocks in the Lampung Districts are largely covered by an older acid tuff deposit. These parts can perhaps best be described as a largely tuff-covered and block-faulted peneplain or piedmont plain. The more northerly parts of this eastern zone are characterized by folding, as will be evident from the description (p 52). Rhyolitic plugs have pierced the acid tuff cover along the fault bordering the Sekampong block to the W, while a basaltic lava outflow occurred further to the E along a fault in the Sukadana area. The following geomorphological units are distinguished by the present author: 1. the Bengkulu Block, 2. the Median Graben, 3. the mountainous areas to the E of the Median Graben, and 4. the low areas of eastern Sumatra.

They will be described in some detail on the following pages.

b. The Bengkulu Block

1. *General features*

The fault scarp marking the northeastern limit of this gently oceanward-dipping block decreases gradually in height from more than 1,000 m in the NW to 50 m and less near Cape Tjina in the SE (Fig. 10). The width of the block also decreases in this direction. This is particularly well illustrated by the sharply pointed southern end of the 50-m contour line near Cape Tjina. The oceanward dip of the block is also clearly visible on Chart 71, Sunda Straits (scale 1 : 200,000). The 100-m depth contour lies rather far off-shore in the S and approaches the coast further N, thus indicating a steeper dip in the N. The seaward edge of the block is marked by a minor submarine ridge, possibly a drowned (pseudo) barrier reef, and is crowned by the small coral island of Betuah. The northern continuation of this ridge is found in the small and slightly elevated coral reefs forming a number of capes along the W coast and thus giving it an irregular shore-line. A sudden increase in depth occurs further seaward, for which Westerveld (1933) supposes the occurrence of a fault scarp approximately parallel to this depth contour.

The surface of the block is a tilted and subsequently dissected peneplain, bordered by Neogene sediments at the seaward side. The southernmost part is almost exclusively composed of these sediments, whereas to the W of the Suoh depression (62 x 8) an older andesitic and dacitic series and a granitic intrusive occur, with only a narrow Neogene zone. Further N, volcanic breccias and tuffs are found with a much broader Neogene belt (Van Tuyn, 1932). Granites are predominant, and form the highest northeastern parts of the block (56 x 8), reaching 1,811 m.

It is remarkable that the Neogene belt is narrowest and lowest (max. 100 m) near the batholith, and reaches heights of 250 to 300 m further to the N and S. One wonders whether the presence of this batholith influenced the more recent uplift of the area. The granite is rather susceptible to erosion: it does not rise from the peneplain like some of the other granites, and is only characterized by its more irregular crest lines. The older andesitic terrain is, however, distinctly more resistant and stands out in relief. The valleys here are narrower and have a steeper and more irregular gradient with waterfalls and rapids, which is in striking contrast with the more open valleys and graded long profiles found in the Neogene and granitic areas. Some isolated andesitic and dacitic hills rise from the Neogene zone and form 'inliers', indicating that these monadnocks of the peneplain were buried in Late Tertiary times and subsequently re-emerged. An example is found NE of Bengkulu.

The older andesitic terrain (Early Neogene) can be distinguished from the Quaternary volcanoes because the products of the latter fill the valleys. The distinction is not always easy to make, however, because the volcanism has probably never completely ceased.

Folding and erosion occurred during the Early Neogene and the transgression should be placed in the Late Neogene, although it did not occur simultaneously everywhere. The marine deposits near Krui, for instance, directly overlie the old andesites, whereas N of this town continental volcanic breccias are intercalated. Since no outcrops of Neogene occur within the peneplain area, it is evident that the latter was formed when the shoreline was located at or near the landward limit of the Neogene zone. It must therefore have been formed after the diastrophism at the beginning of the Neogene. The ultimate uplift and tilting occurred after the subsidence of the coastal areas just mentioned, and are to be correlated with the Plio-Pleistocene diastrophism.

2. *Drainage pattern*

The intensity of the present dissection of this peneplain is, apart from the lithological influences mentioned above, primarily determined by the heights to which the various parts were lifted and by the steepness of the blocks. Since the uplift occurred more or less simultaneously throughout the entire Bengkulu Block, the degree of dissection is no indication of the absolute age of the forms. Fig. 4, showing the contour patterns of a gently elevated zone (sheet III–S. Sumatra 1 : 100,000) and of an intensely uplifted area (sheet IX of same series), illustrates this.

Fig. 4. Two fragments of the 1 : 100,000 topographical map of southern Sumatra showing the influence of elevation on the dissection of the Bengkulu Block. Left: strongly uplifted part (1500 m) W of Lake Ranau and the Simpang Balak River (sheet XXVI). Right: low-lying area (50 m) N of the Bintuhan River (sheet XXI).

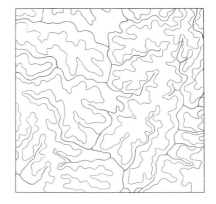

The main rivers of the block have their sources near the fault scarp along the Median Graben, and consequently drain toward the Indian Ocean. They are rather short, due to the narrowness of the block. This situation results in considerable variations in run-off as well as in frequent floods, which render these rivers rather unsuitable for irrigation purposes. They are navigable. The influence of faults on the drainage pattern is not obvious, but can still be traced in several places. Geomorphological indications, for instance, are provided by changes in direction in several rivers. These drainage elements, the location of Tampang and Karangberak bays (Fig. 5), as well as several other relief characteristics, all suggest the occurrence of faults parallel to the Median Graben. Some of these rectilinear drainage elements also extend into the Neogene zone, thus indicating that the fault movements date, at least partially, from the Quaternary.

Another feature of interest is the E-W-striking coastline at the southernmost tip of the Bengkulu Block (Fig. 5). The direction of this coast, which

Fig. 5. Drainage pattern of the southernmost tip of the Bengkulu Block. Drainage lines presumably influenced by structure are found in the dotted zones.

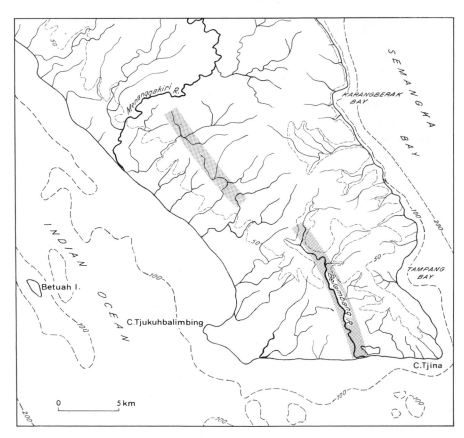

is largely bordered by Neogene sediments, cannot be explained solely by the gradual southeastward submergence of the block. The rivers emptying on the W coast have a consequent direction NE–SW. The Belambang River, located to the E of the sources of these rivers, flows southward, but its tributaries coming from the left again have the consequent NE–SW direction. Whether piracies occurred here cannot be said with certainty, but an influence of fault movements on the drainage pattern and also on the coast line is likely.

Philippi (1916) assumed the occurrence of numerous transverse faults in the block. For many of these, however, geomorphological evidence is very scanty. Indications of faults parallel to the Median Graben, such as abrupt changes of direction of rivers, are generally somewhat more distinct. As a whole, the morphology of the Bengkulu Block is only slightly influenced by these fault movements. The height of the block gradually rises northwestward as far as the Lake Ranau area, but N of Bengkulu it suddenly drops and is there much lower than even in the S. Philippi mentions the occurrence of faults in this area that clearly mark the transition to the intensely block-faulted mountains occurring in this zone further to the N. The existence of faults parallel to the Median Graben is suggested by structurally-controlled drainage elements in several river basins in this northern part of the Bengkulu Block.

The location of the wedge-shaped strip of coastal lowlands marked at its northwestern end by the beautiful offshore bar of Udjungpulau (Udjung = Cape) cannot be explained by riverwork only, and structural factors probably played an important role in its formation. A small horst terminates at the Cape of Bengkulu and is surrounded by lowlands from which rise a number of isolated hills. The Bungkulu Block directly behind and to the N of the town of Bengkulu is very low; a sudden increase in gradient occurs further inland, not far from the Median Graben. A wide zone to the N of Bengkulu is accordingly covered by Neogene sediments. The increase in height of the block to the S of the town is much more regular, and greater heights are therefore reached at a shorter distance from the coast, except in the Mt Sunur area.

3. *Raised beaches*

Other interesting phenomena are the coastal terraces, raised coral reefs, and beach ridges mentioned by Erb (1905). They point to stepwise regression of the sea due to the combined effect of tectonic uplift and Pleistocene changes in sea-level. Such features have been described from various localities NW of Krui at altitudes of 16–20, 25–32, and 45–50 m above sea-level; on their seaward sides they are often bordered by steep cliffs. Some continue inland as river terraces. The occurrence of coral reefs at only 3–5 m above sea-level has already been mentioned.

Nearer to Krui there is a terrace with a height of about 100 m which, according to Van Raalten (1937), decreases to 50 m over a distance of about 15 km in a NW direction. This is considered by Van Bemmelen (1949) to constitute evidence of a greater uplift in the Lake Ranau sector as a consequence of volcano-tectonic up-arching. It appears to the present author that there is a great need for detailed dating and correlation of the numerous remnants of coastal terraces.

Air photos cover parts of the zone of old beach ridges occurring on the lower marine terraces N of the Mana River and near the airfield of Bengkulu. These beach ridges lie parallel to the present coastline or at a slight angle to it. The ridges are remarkably flat-topped and broad, whereas the moats are narrow. Their appearance is therefore completely different from that of the young beach ridges occurring in other alluvial plains. A trellis pattern of minor streams with subsequent sections located in the moats and transverse breaches through the ridges, is characteristic of several of these areas.

A former shoreline, about 2.5 m above mean sea-level, was observed by the author near Bengkulu. It continues in a low terrace bordering the Bangkahulu River. Higher terraces were not observed, but the higher ground (100 m above sea-level) in and near the town probably represents a horst with a marine terrace.

To the N of Bengkulu the Neogene sediments with their marine terraces reach the coast, forming a cliff about 10 m high. The inland continuation of this level is formed by the broad terraces situated 5–7 m above the valley bottom of the Lais and other rivers. Their relative height increases slightly upstream, and reaches 8 m at a distance of 15–20 km from the sea. At this point the height of the peneplain is approximately 150 m above mean sea-level and the river incision is about 30 m deep. There are indications that still higher terraces once existed, but correlation of the various remnants is difficult.

It is noteworthy that several broad and largely dry valleys near the shore debouch approximately 1 m above mean sea-level. This cannot be the result of recession of the cliff coast under the influence of abrasion, since the gradient of these valleys is extremely small. A continued, sub-Recent, relative rise of the coastal zone, which also produced the 2.5 m high terrace at Bengkulu, is thus evident. The area SE of Bengkulu probably underwent a slightly stronger uplift than the surroundings of the Lais River, as shown by the occurrence of a 12 m terrace near Tais. This level can probably be correlated with the previously-mentioned, approximately 10 m high beach ridge area near the Bengkulu airfield and with the 10 m level of Lais, where the rivers are incised to only about 6 m. Tais is situated in one of the few alluvial plains of any importance bordering the Bengkulu Block to the W, extending southeastward from Bengkulu, as indicated on the geomorphological map. There is no major river to account for this lowland strip, and it may therefore be assumed to be composed of marinee sediments which have emerged, as already mentioned, due to structural causes.

4. Volcanoes

The volcanic activity within the Bengkulu Block deserves further attention. It has already been stated (p 25) that the volcanism has probably never completely ceased since the Early Neogene. The Late Neogene volcanoes are in some places so strongly eroded that they can hardly be distinguished morphologically from the Old Andesites forming their base. It is therefore not surprising that they have been confused with the andesites. Moerman (1915) mentions the widespread occurrence of Old Andesites, but the more recent investigations by Westerveld (Ubachs, 1941) seem to indicate that many andesites intrude into the Neogene and that Old Andesites occur seldom or not at all. Steep-sided andesitic diatremas are conspicuous features in the hinterland of Bengkulu. These are relics, 'volcanic necks' of Late Neogene age (Philippi, 1916). The field sketches (Figs. 6 and 7) show two of these necks as well as a number of younger volcanic bodies.

Fig. 6. Field sketch of the tilted Bengkulu Block and the volcanoes crowning it near the Median Graben, as seen from the town of Bengkulu, looking E and N. Note the Mt Bungkuk volcanic neck. The Kaba volcano is located on the western side of the graben.

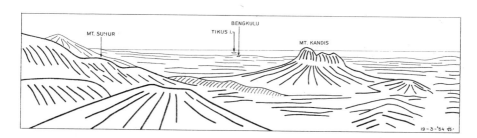

Fig. 7. Field sketch showing the Bengkulu Block and the Mt Kandis volcanic neck as seen from the Mt Kambing pass W of the town of Kepahiang looking SW and W.

The Quaternary volcanic activity was largely limited to the fault scarp bordering the Median Graben, but several volcanoes also rose elsewhere. The locations of the latter are probably influenced by parallel faults. Of the Quaternary volcanoes, Mt Pugung (1,964 m) located SW of Lake Ranau (Fig. 10) should be mentioned. This mountain is still a distinct volcanic cone, in contrast to the volcanoes (58 x 8) located further to the NW near the Median Graben, all of which, however, still stand out clearly in relief.

Another, even more interesting, volcanic area is situated to the N of Beng-

Photo 3. The gentle lower slopes of the volcanoes of Hulupalik (beyond the right edge of the photograph), Hulukokai (on the right), and Tiga (in the distance) stretch far to the W and overlie extensive parts of the Bengkulu Block with volcanic tuffs.

kulu (Photo 3; Figs. 6 and 7), with the Hulupalik-Daun complex as its southernmost volcano (Fig. 8). The Daun products probably blocked the Ketaun River in the Median Graben (p 39), and the gentle slopes of this complex spread far westward, covering the peneplain of the Bengkulu Block. The Kemumu resettlement area is located here (Fig. 8). The Hulupalik has a distinct horseshoe-shaped crater rim, as indicated on the geomorphological map.

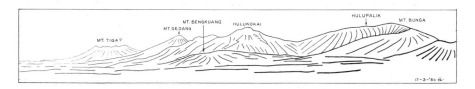

Fig. 8. Field sketch of the volcanoes of the Bengkulu Block, looking N from Gelagah hill in the Kemumu area.

The volcanic range extends to the NW, parallel to the Median Graben, and displays horseshoe-shaped old craters, hot springs, and volcanic fans. At the northern end two small cumulo-domes are located near the fault bordering the Median Graben to the W (Photo 4).

Between the two volcanic complexes of the Bengkulu Block, lying some 160 km apart, the geomorphological map indicates two other fan deposits (56.5 x 7.5 and 53.5 x 7.5). Although they strongly resemble fluvio-volcanic fans, no known Quaternary volcanoes account for them.

c. The Median Graben

1. *General features*

This conspicuous and complex feature stretches from Semangko Bay in the S to the Island of Weh in the N and marks the crest of the Barisan geanticline. E of it, a belt of complex mountains (see geomorphological map) occurs in southern Sumatra, and it is likely that prior to the ultimate up-arching of the Barisan Mountains the main divide was partly located there. These mountains, according to Westerveld (1953), form part of the Late Cretaceous or Early Eocene Sumatra orogene, which further N continuously forms the backbone of the Barisan Mountains. The subsequent changes in the drainage system under the influence of tectonic movements and the related volcanism were so fundamental that a reconstruction of the original pattern poses almost unsurmountable difficulties. A number of interesting drainage problems, wind gaps, and possible piracies are mentioned in this paper, but much remains obscure.

At present, parts of the Median Graben are drained towards the Indian Ocean and other sections towards the Straits of Malacca. At some localities the main divide is therefore located in the graben area and is low or even almost imperceptible. Since the differential tectonic movements in the Barisan Mountains have continued up to the present, the drainage of the Median Graben at several localities is hampered, which limits full use of the agricultural and other potentialities of these intramontane plains.

2. *Semangka Section*

Semangka Bay is typically wedge-shaped, and a triangular alluvial plain of the river of the same name occurs in its narrow NW end. It is noteworthy that the fault scarp bordering the bay to the W is straight and runs in the

Photo 4. Details of the western fault scarp of the Median Graben near Talang Leak, S of Muaraaman, in southern Sumatra. The extinct stratovolcano Mt Lumut is seen at the left, and two cumulodomes, locally called *sepikul*, are seen along the fault, near the right edge of the photograph.

direction of Sumatra. This scarp recedes only slightly and stepwise near its southern end, probably due to faulting (Fig. 10). The eastern fault scarp, on the contrary, is curved outward. Alternating capes and bays occur in the Old Andesites cropping out here, indicating submergence of its southernmost part. The island of Legundi and some islands further E apparently were separated from Sumatra as a result of this subsidence. The Semangka plain narrows upstream and finds it continuation in the valley of the Semangka River running parallel to the western fault scarp. E of it there is, however, a parallel river (the Semung), possibly indicating the eastern scarp buried below andesitic tuffs.

The contrast between the straight western scarp and the curved eastern scarp becomes less pronounced if one takes into consideration the 200 m depth contour, which at either side of the graben deviates considerably from the Sumatra direction. The horst of Tubuan Island, narrowing toward the NW, occurs in the middle of the bay. The products of the young basaltic Tanggamus volcano reach the bay directly to the SE of Kotaagung and cover the Old Andesites of the fault scarp (Photo 5). Consequently, the terrain to the E of the Semangka alluvial plain rises gradually, whereas the western fault scarp there is high and steep.

Following the Median Graben toward the NW, we arrive at the Suoh depression (62 x 8) and the smaller adjacent depressions of (62 x 8) Tikarberak and Antatai (63 x 8), where transverse faults give the graben a complex character (Fig. 10). These depressions have a straight northeastern border following the continuation of the Semung River mentioned above. These depressions should be considered as 'volcano-tectonic' only in the limited sense that acid (dacitic) lava outflows are associated with the faults bordering them. Such outflows also occurred over a considerable distance in the

Photo 5. Photograph taken looking E from the coast of the Semangka Bay (southern Sumatra) near Kotaagung. In the foreground, the volcanic products of the slightly dissected Tanggamus stratovolcano (left) reach the bay. The fault scarp (in old andesites) bordering the bay to the E can be seen in the distance at the right.

Photo 6. The northwestern area of the Suoh depression, forming part of the Median Graben of southern Sumatra. The forested ridge in the background represents the western fault scarp. This area was the scene of a phreatic eruption in 1933, which gave rise to the barren hill on the right (Mt Ratus), the fumaroles in the distance, and the lake.

Median Graben NW of Antatai, as a result of which there is no alluvial plain. Products of the andesitic stratovolcanoes often cover these acid lavas, and their age is therefore greater than at least some of the latter volcanoes. No tuff or ignimbrite eruptions are reported for these depressions, but the northwestern corner of the Suoh depression was the scene of a phreatic eruption in 1933 (Stehn, 1934). Photo 6 shows the main eruption centre and the additional products ejected by the vents along the northwestern fault scarp. The shallow lake in the foreground was formed at the same time.

3. *Ranau Section*

It is possible that dacitic lava outflows also occurred further N in the graben but were buried by the tuff/ignimbrites of the Ranau eruption. These thick deposits, the eruption centre of which was a depression measuring approximately 16 x 12 km and containing lake Ranau, extend in the graben to the SE, where the Semangka River is located, and also in the northwestern graben area drained by the Kuala River. An important ignimbrite plateau having a gradient of about one degree stretches to the NE in the direction of Muaradua. Deep (> 200 m) canyons are formed in the acid material, which is unstratified and therefore was probably deposited during one catastrophic eruption. The products even reached the peneplain of eastern Sumatra by way of the Komering valley (see geomorphological map). All the surrounding Quaternary stratovolcanoes are older than the Ranau eruption; only the well-preserved cone of the Seminung volcano, located on the southeastern shore of the lake, is younger (Photo 7). Lava-flows from this volcano are clearly traceable underneath the vegetative cover. The southeastern ignimbrite flow is separated from the lake by this volcano, and it also lies at a considerably higher level (900 as against 600 m). This higher position ex-

Photo 7. The S shore of Lake Ranau (southern Sumatra) with the extinct Seminung volcano (left) as seen looking W from Tandjungdjati. Note the delta formation in the left foreground and the western fault scarp of the Median Graben in the distance on the right.

cludes a genesis by volcanic mud flows, but is quite normal for ignimbrites, also in other parts of the island. There can be little doubt, in the author's opinion, that all the 'volcano-tectonic' depressions located in the Median Graben existed prior to the ignimbrite eruptions and other acid volcanic activities (Verstappen, 1961).

The scarp surrounding the lake in the W and S is hundreds of metres high and intercepts the straight course of the western fault scarp of the Semangka rift. This western fault scarp is certainly complex in nature. A distinct sliver occurring near Liwa (61 x 8) has a pronounced influence on the relief and drainage characteristics of this area. This is illustrated by Fig. 9, which shows

Fig. 9. Fragment of the 1 : 100,000 topographical map of southern Sumatra (sheet XVIII) showing the fault scarp (WF) bordering the Median Graben on the W near Liwa and a sliver occurring there (right part of figure). Note the influence of the fault scarp and the secondary fault (SF) on the drainage.

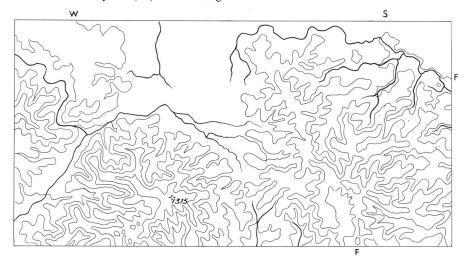

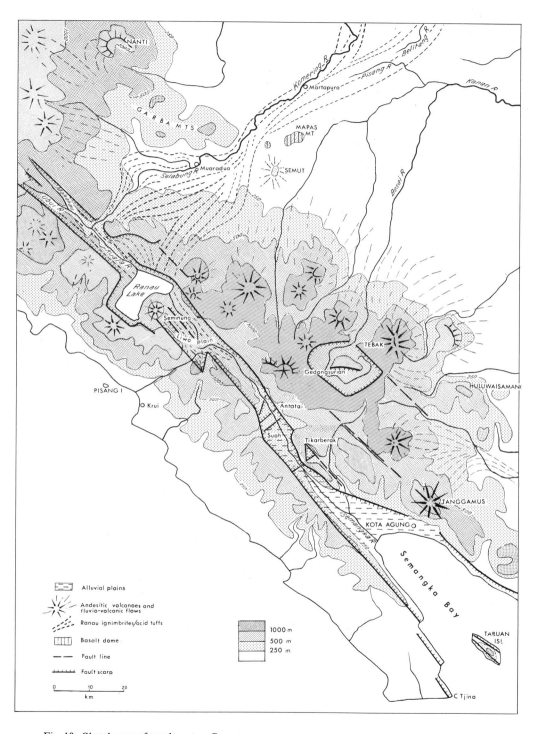

Fig. 10. Sketch map of southwestern Sumatra.

a fragment of sheet XVIII of the 1 : 100,000 topographical map of southern Sumatra, WF indicating the main western fault scarp and SF the fault bordering the sliver.

Another small horst, Mt Uluhun, occurs in the graben directly to the SE of the Seminung volcano. Van Bemmelen (1932 a) mentions the northeastward shift of the divide between the Semangka River and the Warkuk River draining towards Lake Ranau, as a result of the Ranau eruption. The latter river is not a true underfit river, since it flows in a broad graben and not in a valley carved by its own action. Near the lake, the Seminung volcano forces the river to cut through andesites, where it forms a narrow, 150 m deep gorge.

The history of the drainage system is certainly more complicated, however. The upper reaches of the Laki River, which discharges into the Indian Ocean N of Krui, are located in the Median Graben near Liwa (Fig. 10). The small stream breaks through the enormous western fault scarp and traverses the Bengkulu Block before it reaches the Indian Ocean. The rather steep gradient of this part of Bengkulu Block means that the Laki River is a potential pirate of the rivers flowing in this part of the graben. No terraces could be observed along this river in the Bengkulu Block, but a distinct knickpoint occurs in the valley sides. The origin of the breach in the western fault scarp is obscure, however. Comparable cases are known to the author from central and northern Sumatra. The possibility that here the Median Graben once was drained toward the Indian Ocean cannot be excluded.

The eastern fault scarp decreases in height near the lake. Ignimbrites crop out along the NE lake shore, but only reach a height of 700 m. A higher escarpment, where there is an outcrop of older rocks, occurs at some distance from the lake and can be followed toward the NW, where it represents the fault scarp bordering the Median Graben on the E. The existence of the Semangka fault to the N of Lake Ranau was proved by Van Bemmelen, who observed a throw of 500 m.

4. *Mekakau-Tandjungsakti Section*

Lake Ranau, at 540 m above mean sea-level (Depth 22 m), is drained to the NW by the Kuala River (Fig. 11). Before this river breaks through the eastern fault scarp it is joined by the Mekakau River coming from the opposite direction, and together they form the Selabung River. A local horst, Mt Asadimana (58 x 8.5) occurs alongside the upper Mekakau River (Fig. 10). NW of this point, the Median Graben becomes ill-defined over a distance of some 60 km, and no escarpments can be distinguished. The divide shifts to the E and runs over an old Volcanic cone (2,819 m; see geomorphological map).

A broad and well-developed part of the graben is formed by the plain of Tandjungsakti (55.5 x 8.5), SE of the Dempo volcano. This area is drained toward the Indian Ocean by the Mana River in a zone where Old Tertiary sediments of the Bengkulu Block reach the Median Graben, the granitic zone being interrupted here

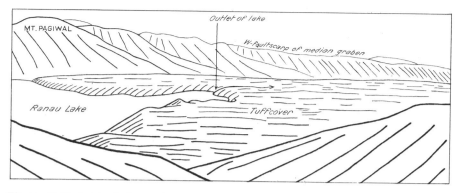

Fig. 11. Field sketch of the northeastern part of Lake Ranau near its outlet located in the Median Graben, which is filled with tuffs of the Ranau eruption.

The Median Graben appears to have a complex nature in this area (Fig. 12). Two headwaters of the Mana River flow to the NW, and then this river breaks through what seams to be a minor horst to turn into the Tandjungsakti plain. There it is joined by two parallel rivers coming from the opposite direction, and the combined river breaks in its turn through the Bengkulu Block.

5. Keruh-Musi Section

The Keruh River flows northwestward in the graben, joins (53 x 8.5) the upper Musi coming from the opposite direction, and then breaks through the eastern fault scarp. The graben is well developed in this area and increases in width from the upper Keruh northwestward (Fig. 12). The western fault scarp is consistently high. The eastern fault scarp reaches a height of about 1,000 m in the S and is considerably lower (about 450 m) where the Musi breaks through it. A NW tilt of the block to the E of the graben accounts for this situation. A local horst, Mt Kulitmanis-Gelega (714 m), is situated to the W of the upper Musi near its confluence with the Keruh River. The eastern fault scarp N of the confluence is low and ill-defined: the slopes of the active Kaba volcano enter the graben, and it is only at the other side of this volcano that a distinct eastern fault scarp reappears, although shifted to the NE. This may be due to a transverse fault, a view supported by the alignment of the Kaba craters perpendicular to the Median Graben.

6. Ketaun Section

Further on, the Median Graben becomes narrower and is complicated by some local horsts. It is drained by the Ketaun River, which flows to the NW and finally breaks through the western fault scarp to reach the Indian Ocean. Its upper course in the eastern mountain block, however, runs to the S and suddenly turns to the NW where it enters the graben. This suggests that the

Fig. 12. Sketch map of the western part of southern Sumatra.

upper Ketaun River formerly drained toward the Musi River. Actually, the divide between these two river systems is almost imperceptible, being located in a swamp (Fig. 12). An alluvial plain along the upper course of the Ketaun River and the occurrence of a terrace deposit at 950 m, where it enters the graben, point to a ponding-up of the Ketaun River. This deposit is mainly volcanic and originates from the western volcanoes (Daun, Hulupalik).

The narrow, winding Lake of Tes has been explained by Moerman (1916) as the result of drowning of the valley, due to faulting. He also mentions several faults near Muaraäman. Various faults and a major landslide in this part of the graben can indeed be recognized from the air.

The next part of the Median Graben, draining through the Seblat River toward the Indian Ocean, becomes so narrow that one wonders whether it is still a true graben or whether the longitudinal depression resulted from erosion along a single fault line.

d. The mountainous areas to the East of the Median Graben

1. *General features*

These areas form the eastern flank of the Barisan geanticline. A number of eastward-tilted and horizontal blocks can be observed; this 'pre-volcanic surface' forms the basis of the Quaternary volcanoes. The nature and age of this surface is presumably the same as that of the Bengkulu Block to the W of the Median Graben. It is usually intensely dissected, so that little or nothing of it remains in its strongly uplifted parts (e.g. in central and northern Sumatra). Complex mountains largely composed of Pre-Tertiary sediments and of

Fig. 13. Two fragments of the 1 : 100,000 topographical map of southern Sumatra showing the difference in dissection of recent and older volcanic slopes. Left: strongly dissected southern slope of the Ratai volcano (sheet IX). Right: weak dissection of the southern slope of the young Tanggamus volcano (sheet III).

Photo 8. Vertical aerial photograph of the deeply dissected extinct Butung volcano located SE of Mt Dempo, southern Sumatra.

igneous and metamorphic rocks, also occur in this zone. Their height is usually somewhat greater due to resistance to erosion and/or stronger uplift, and they therefore have a great influence on the outflow of volcanic products toward the eastern peneplain.

The largely andesitic stratovolcanoes show striking differences in the dissection of their slopes, because their activities did not cease simultaneously. A number of them are still active. Fig. 13 shows two fragments of the 1 : 100,000 topographical map of southern Sumatra, one depicting the southern slope of the Ratai volcano (left) and the other that of the Tanggamus volcano (right). It is evident from the shallower dissection that the activity of the latter volcano ended more recently. Photo 8 is a vertical aerial view of a still older volcano; thee deep dissection of its slopes is proof that this volcano has been extinct since tthe Middlde Pleistocene or earlier. It must be kept in mind that the degree of dissection is a measure of the time elapsed since the volcanoes became extinct, but has no bearing on their time of origin.

2. *Semangka and Ratai Blocks* (Fig. 14)

The Semangka Block (Van Bemmelen, 1949) situated E of Semangka Bay, is largely composed of Old Andesites. It reaches heights of 700 m near its western edge and slopes downward gradually toward the NE. Some higher tops rise from its intensely dissected surface. Subsidence of the southern part, as mentioned on p 32, accounts for the numerous capes and bays along the southern shores and for the islands separated from the main block. The block is drained by consequent rivers, and no influence of structure on the drainage pattern can be observed. The more intense dissection of this block and the position of its divide at considerable distance from Semangka Bay point to an earlier emergence of this block as compared to the Bengkulu Block, where Neogene outcrops are extensive and where recent uplift occurs.

A more or less parallel fault scarp of lesser importance is found E of Ratai Bay (see e.g. the depth contours of chart 94 to scale 1 : 75,000). The andesitic Ratai volcano, which has a horseshoe-shaped crater rim, is located on this fault and covers the greater part of the scarp.

Between the Semangko and Ratai Blocks and S of the Ratai volcano there

Fig. 14. Sketch map of the southeastern part of southern Sumatra.

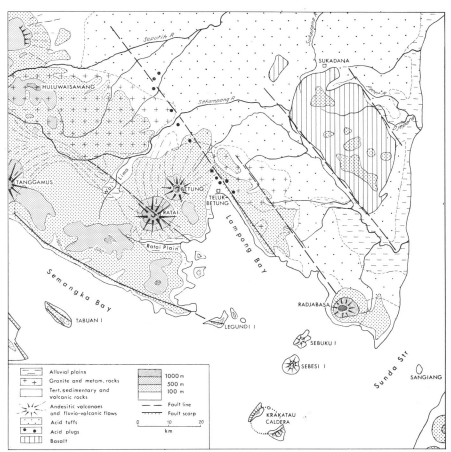

is an alluvial plain with a subsequent river collecting the consequent streams of the Semangka Block and the radial streams of the volcano.

Another alluvial plain, the Wai Lima depression (65 x 9.5), occurs within the same low zone to the NW of the Ratai volcano. The lowest part of the Wai Lima plain is formed by a swamp from which rise several isolated andesitic hills. It seems logical to conclude that the drainage of this zone was directed toward Ratai Bay before the birth of the Ratai volcano, but has since been diverted to the E. It is now drained by a tributary of the large Sekampong River.

There is a project under consideration to turn the swamp into a reservoir for irrigating a larger area in the Metro resettlement area along the lower Sekampong. The disadvantage of this project is that due to the shallowness of the swamp, it will cost the flooding of a large, fertile, and cultivated area to achieve the irrigation of the rather poor soils in the low areas further to the E (p 48).

3. *Hulawaisamang granites and surrounding volcanoes*
One of the largest granitic and metamorphic areas of Sumatra is found N of the Wai Lima plain. Part of this area is base-levelled, although less so than the islands of Bangka and Belitung, but another part bears the residual Hulawaisamang Mountains, rising up to 1,080 m and topped by some Tertiary sediments. Quartz grains are abundant at the surface of the soils. Sand and pebbles originating from the granites and gneisses occur extensively on top of the dacitic tuff plains further E, indicating that they were deposited by more competent transporting agents than the present rivers (Westerveld 1931). The possibility that this coarser material represents a period of less intense chemical weathering should also be kept in mind.

The granitic and gneissic Hulawaisamang Mountains influenced the distribution of the fluvio-volcanic fans extending from the surrounding volcanoes (Fig. 14). In the S, the lower slopes of the Tanggamus could spread freely toward the NE until the fluvio-volcanic flows were deflected toward the SE by the hills in the metamorphic area. Some flows may have extended over the lowest base-levelled part of the crystalline rocks, where remnants of andesitic tuffs are preserved. Further downstream, the Sekampong River was pushed toward the basement complex by the flows from the lower slopes of the Ratai-Betung volcanoes.

The same is the case along the northern border of the granitic area, where the Seputih River was pushed against the granitic mountains by the fluvio-volcanic fans coming down from another group of Quaternary volcanoes.

4. *Gedongsurion graben and adjacent block mountains*
The continuation of the Semangka Block occurs to the NE of the Tanggamus volcano and there forms a plateau with an altitude of 1,100–1,300 m, partly covered by products of the Quaternary andesitic volcanoes.

The culminating part contains the sunken volcano-tectonic Gedongsurion graben. It may, according to Van Bemmelen (1933, 1949), have been one of the eruption centres of the extensive acid Lampung tuff sheet. Later, it produced some acid lava along a fault.

The depression is separated from the Median Graben by some extinct andesitic volcanoes and, also to the NW, by an extensive volcanic area with numerous extinct cones. The block is buried here under volcanic products, and reappears only to the N of Lake Ranau. These products spread freely eastward to the low areas of eastern Sumatra, covering the dacitic tuffs (ignimbrites) with andesites. The transition between the fertile soils on the latter and the acid areas is morphologically almost imperceptible.

5. *Komering Gap and Garba Mountains*

The ignimbrites of the Ranau eruption flowed eastward through the 'Komering Gap' between the above-mentioned stratovolcanoes and the complex Garba Mountains further to the NW. The ignimbrites are evidently younger than the andesitic volcanoes. One apex of the ignimbrite fan is formed by the site where the Silabung River, which drains Lake Ranau, breaks through the Barisan range, another is situated directly N of Lake Ranau.

Lower down in the gap there are some peculiar volcanic eruption centres: the acid fissure eruption of the Pematang Semut and the sub-recent andesitic lava domes of the Mapas Mountains both described by Van Bemmelen (1931, 1949).

The complex Garba Mountains gradually decrease in height towards the SE and disappear under the Ranau ignimbrites. It is noteworthy that the upper Komering branches flowing toward the 'gap' are forced against the western limit of the Garba Mountains by the volcanic flows. The Komering in the gap cuts a narrow gorge (60 x 10) through the Garba granites instead of continuing its course in the ignimbrites, in which it has formed a steep-sided valley further upstream. In this gorge, terraces occur.

The Garba area was almost completely submerged in the Early Neogene and its present position is due to later orogenic uplift. A similar situation exists further to the NW, where the Musi River traverses the complex Gumai Mountains.

Another extensive area of Quaternary andesitic stratovolcanoes occurs between the Garba and Gumai mountains. The cone of the extinct Mt Nanti (1,619) m) with a horseshoe-shaped crater rim open to the SE, partly covers the northern end of the Garba Mountains. Other old cones can also be traced. The pre-volcanic surface is deeply buried in this volcanic complex and only appears further to the W, where an extensive fluvio-volcanic fan spreads eastward through the gap of Sugiwaras (57 x 10.5) (Fig. 17). Two parallel rivers flow through this gap, one of which, the Enim River, turns N after having passed the gap; the other, the Ogan River, turns E. Between them is situated a huge fan with a radial drainage, where the andesite tuff flows spread out over the eastern peneplain.

Traces of the pre-volcanic Barisan surface, at a height of about 1,400 m, can be seen in the midst of the volcanic area along the upper courses of the Enim River.

6. *Pasemah and Gumai Mountains*

In the next section, the Pasemah Highlands, the pre-volcanic surface must have been considerably lower, because this plateau has an altitude of approximately 700 m and is, moreover, covered with a thick layer of ignimbrites and andesitic tuffs. This difference in height may be due to transverse faults; there is, at any rate, a low and wide gap here, connecting the Pasemah plateau with the peneplain in the E and followed by the Lematang River.

The Pasemah ignimbrites, dating from the Late Quaternary, probably originated from fissure eruptions. They are largely covered by the Dempo volcanic products, also in the depression S of the Gumai Mountains, where steep valleys, 100–200 m deep, occur. The stereogram in Photo 9 illustrates

Photo 9. Stereogram of part of the Pasemah ignimbrite plateau near Pelangkendidai, SE of Pagaralam (southern Sumatra); scale 1 : 29,000. The Lematang River is deeply incised into the plateau, occupying a steep to vertical valley with gentle upper slopes.

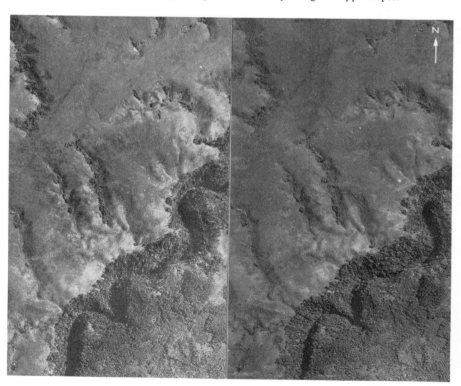

Photo 10. The northern Pasemah area (*Lintang*) with the Dempo volcano in the background, as seen from Talang Mesundang in the NW (southern Sumatra).

the permeability of the ignimbrites. Two levels can be distinguished here, but there is no indication of the occurrence of two tuff or ignimbrite flows in the much lower area to the N of the Dempo Volcano (Photo 10), where the rivers are less deeply incised. The recent products of the active Dempo volcano spread far to the NW in this area as a result of a breach in the volcano on that side. A distinctly radial drainage pattern has formed there.

The Pasemah Highlands (54/56 x 9) form a longitudinal depression zone between the eastward-tilted Barisan Blocks, partly buried below the Dempo volcano, and the complex Gumai Mountains. The Musi River, originating in the Median Graben, breaks through the lowest parts of both the Barisan and the Gumai mountains. It is noteworthy that the Pasemah is actually one of a series of depressions that can be followed southeastward to the fluvio-volcanic slopes W of the Garba Mountains (58/59 x 9.5), the Gedongsurion depression (62.5 x 9), the Wai Lima plain (65 x 9,5), and Ratai Bay (66 x 9.5). Thus, this depression zone is interrupted, and at least the two depressions along the Gumai and Garba mountains bend eastward in wide gaps towards the eastern lowlands.

The complex and rather asymmetric Gumai Mountains, located to the E of the Pasemah Highlands, can be regarded as a large anticline of the Neogene mantle; below the Neogene unconformity the Mesozoic is more strongly folded. The culminating ridge is not, however, found in the tectonic axis but in the Neogene of the southern flank, consisting of andesitic tuffs. The main axis decreases in height toward the NW, where the Musi River traverses the mountains in a narrow gorge. This is probably antecedent in origin, but no distinct terraces could be traced by the present author. Tobler (1925) assumes a transverse fault at this locality. The drainage pattern of the Gumai Mountains (Fig. 15) was studied by Tissot van Patot (1919), who states that subsequent rivers forming a trellis pattern are common. Various karst phe-

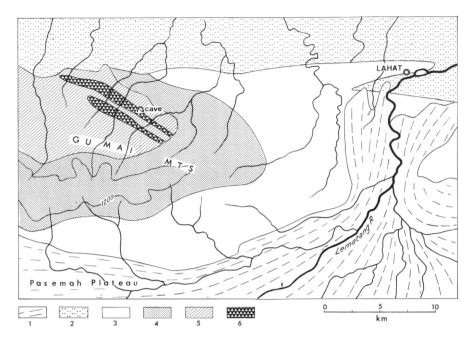

Fig. 15. Drainage pattern and geology of the easternmost part of the Gumai Mountains, simplified after Musper (Geological map of Sumatra 1 : 200,000, sheet 16). The area can be considered an eastward-plunging asymmetric anticline with a steep to overturned northern flank.

Key: 1. Andesitic fluvio-volcanic flows
2. Palembang beds
3. Telisa beds
4. Lahat series
5. Saling and Lingsing series
6. Limestone ridges

nomena (caves, underground rivers, dry valleys) can be observed in the Mesozoic limestones in the northern part of these mountains (Musper, 1934).

The northeastern limit of the Gumai Mountains is well defined, but N of the Musi gorge the asymmetric main range passes more gradually into the eastern peneplain. A few smalll rivers to the W of this range, flowing in the NW–SE (Sumatra) direction, probably as a result of faulting, approximately indicate the transition to the eastward-titled Barisan Block mountains. This block disappears further to the NW under the active Kaba stratovolcano.

7. *Kaba volcano*

This is actually a twin volcano whose older northwestern cone is extinct. Hot springs occur on its eastern side. The several craters of the Kaba are located on a WSW–ENE-directed line, probably a transverse fault, to the NW of which the Median Graben suddenly widens, as mentioned on p 37. The

Photo 11. Top area of the Kaba volcano (southern Sumatra) on March 25, 1954, looking E from the northern edge of the Kaba Lama crater. The Kaba Baru crater appears in the foreground. Note the intense gullying in the loose ashes deposited during the eruption of 1951. The ash cone in the middle distance holds the Vogelsang crater. A small lake can be seen in the foreground.

oldest crater, i.e. the one to the SSW, and the central one, are in the fumarolic stage, but the northeastern or Vogelsang crater, formed during the 1873–1892 eruption period, is still active.

Photo 11 and Fig. 16, dating from March 1954, depict the summit area and

Fig. 16. Field sketch of the top area of the Kaba volcano made on March 25, 1954, looking eastward from the rim of the Kawah Lama crater. The cone in the centre contains the Vogelsang crater.

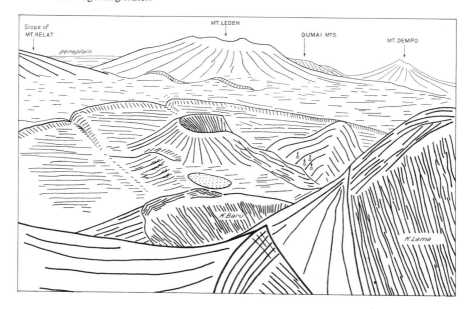

show the effect of the 1951 eruption. Vague indications of a large old caldera can be distinguished. The lahar products of the Kaba and a few older volcanoes N of it flow off freely to the E, toward the plains, through a wide gap N of the place where the Gumai Mountains end. To the NW of this breach there is a large complex mountainous area extending into central Sumatra, and signs of volcanic activity become less evident.

e. The low areas of eastern Sumatra

1. *Sekampong Block*

The southernmost tip of Sumatra, E of Lampung Bay, has the distinct characteristics of an eastward-tilted block mountain, though it is much lower than the Semangka and Ratai Blocks mentioned earlier (p 23). Van Bemmelen (1949) applied the name Sekampong Block to this region. A maximum height of about 200 m is reached near Lampung Bay, and the terrain gradually becomes lower toward the NE. Only a number of more recent dacitic and andesitic eruption points along the western fault scarp emerge from this level.

The surface of the block is largely base-levelled, and also granitic areas, such as an intrusive body E of Telukbetung, are morphologically almost unconspicuous. It seems to the author that this region belongs to a peneplain most likely forming the extension of the one described from the Bengkulu Block and the block mountains E of the Median Graben. The peneplain, however, is completely covered by dacitic Lampung tuffs, and the present land forms can thus best be described as a tuff plateau (Photo 12) that underwent some block faulting. The tuffs probably originated from fissure eruptions along one or more faults in the southern part of the Lampung Dis-

Photo 12. The tuff-covered Lampung Block with the Tanggamus volcano in the background, seen looking SW from the Metro area of southern Sumatra.

Photo 13. The tuff-covered Lampung Block N of Telukbetung with the Kedaton tholoid in the foreground (right) and the extinct Betung stratovolcano in the distance, seen looking W from the top of the Sulah tholoid.

tricts, the cover decreasing in thickness toward the NE. This origin is comparable to that of other tuff/ignimbrites occurring in various parts of the island.

No fault scarp marks the northwestward (inland) extension of the Lampung Fault, but a number of acid (liparitic, dacitic) boccas or tholoids indicate its location. Photo 13 shows one of these steep-sided hills formed by the outflow of viscous lava after the deposition of the Lampung tuffs.

Structurally controlled drainage elements (66/67 x 10.5) indicate the presence of another fault line to the E of, and parallel to, this main fault scarp along Lampung Bay. The southeastern extension of this second fault line approaches the coast of the bay, and the extinct Radjabasa volcano is located here. This is actually a twin volcano, the most eroded part of which is situated to the SE. Continuing along this line to the SE we find a considerably older and lower andesitic country having more relief than the acid tuff plain. It is noteworthy that the resettlement areas on these andesites are considerably more flourishing than those located on the dacitic tuffs.

Various coastal features, such as deep bays and small islands, indicate submergence of the coast. A slightly elevated (2 m) coral reef near Mt Radjabasa points to a minor sub-recent regression of the sea, however.

Several volcanic islands occur in the Sunda Straits. One of the relics of the old Sangian caldera is situated on a line running parallel to the longitudinal Lampung fault and through the Radjabasa volcano to the SE. The others are arranged on a line perpendicular to the main faults, and comprise the severely eroded islands of Sebuku and Sebesi and the ill-famed Krakatau volcano. The effects of the Krakatau eruption of 1883 are still traceable in Sumatra. Large black coral boulders swept along by tsunami accompanying the eruption, are a frequent occurrence along the shore above mean sealevel. The distribution of the Krakatau ashes of 1883 ($>$ 5 cm thick) is indicated on the geomorphological map.

2. *Sukadana Basalt plateau*

The drainage pattern of the areas further to the E reveals a number of interesting geomorphological phenomena. The Sekampong River, on emerging from the Barisan Mountains, takes a consequent easterly direction until it reaches the basaltic Sukadana lava plateau. From there it follows a rectilinear course in a southeastern (Sumatra) direction flowing along the straight edge of the lava plateau. It collects several rivers from the dacitic tuff plains to the W, but no branches come from the recent and porous basalts. Philippi (1916) mentions the basalt outflows as a possible cause of this situation, but since the Sekampong River maintains its straight SE direction downstream from the basalts as well, it is evident that this course is structurally controlled. Van Tuyn (1931) states that the quartzitic sands and pebbles underlying the basalt are situated at a lower level near the Sekampong than further to the E. Consequently, the Lampung tuffs between the basalt plateau and the smaller isolated patch further to the NE reach a height of 48 m above sealevel, whereas further W the altitude is only 33 m. The existence of a fault along the river, already supposed by Van Tuyn and also by Philippi, therefore seems likely.

A similar situation is found at the other side of the basalt area, where there are some indications that tectonic movements favoured the origin of the drainage lines stretching in a NNW–SSE direction, parallel to the edge of the basalts, where the Sukadana River is deflected northwestward and where the Djepara and Penet rivers flow southeastward in broad swampy depressions. Structural control of the drainage seems probable. It is therefore likely that a consequent drainage system originally existed in the southern Lampung Districts, a view supported by the occurrence in many localities of quartzitic sands and pebbles on the dacitic tuffs. The basalt outflows are probably related to, and thus occurred at the same time or after, the faulting movements.

Subterranean drainage is important in the basalt area, and because the underlying dacitic tuffs are considerably less pervious, the water emerges at the edges of the basalt plateau, forming numerous springs. The surface is slightly undulating, and gentle depressions with swamps and small lakes fed by springs occur between the eruption points. Lake Djepara (66.7 x 12.3) is the largest of the lakes. Basalt outcrops are frequent along this lake, and stratified tuffs also occur.

A delta is formed where a small river enters the lake. The soils are very fertile, but water has to be obtained by pumping. There are possibilities for the construction of a dam downstream from Lake Djepara, though leakage is a problem (Hetzel, 1939).

3. *The narrow alluvial plain in the South*

The alluvial plain bordering the E coast is remarkably narrow in the S as compared to the broad flats occurring further to the NE. Evidently, the width of the alluvial plain is not primarily determined by the rate of sedimentation of the rivers, but must be largely governed by recent tectonics resulting in subsidence and upheaval. It is noteworthy that sandy/loamy beach ridges are a common occurrence at short distances from the hinterland. Such ridges rarely occur in the vast alluvial plains further N, but they are a dominant feature in the narrow alluvial strip in the S under discussion. The landward edge of this strip lies at approximately 5–10 m above sea-level and represents the effect of a sub-recent lowering of the sea-level or rise of the land. The lower courses of the rivers in this area and also further to the N are astonishingly deep which is related to the dominance of fine-textured sediments in the plain resulting from the intense chemical weathering of the upstream areas in this humid tropical environment. The lower courses form estuaries, a fact explained by Weeda (Baartmans *et al.*, 1947) as due to the rather large vertical tidal range. The N–S direction of the coastline bordering this part of Sumatra is certainly an anomaly if one considers the general NW/SE trend characterizing the structure of the island. Whether its position is related to a fault structure or to tilting resulting in the partial submergence of the above-mentioned blocks, cannot be said with certainty. There can be little doubt, in the present author's opinion, concerning a structural control of this coastline, which probably also caused the narrowness of the alluvial plain.

4. *Palembang Lowlands, general features*

The drainage of the Lampung tuff plateau further to the NE does not show structural control, which seems to indicate a decrease in the intensity of faulting in this direction. Most of the rivers in these interior lowlands flow to the NE, which is consequent with respect to the Barisan chain. It is interesting with regard to irrigation to mention that these rivers carry less silt than the Sekampong River. Folding does not occur in these areas, which are therefore included on the map in the tuff-covered low block mountains of the Lampung Districts. The gradual transition to the folded geosyncline of eastern Sumatra is tentatively indicated along the Kanan-Tulangbawang River. This limit coincides approximately with the northern limit of the Lampung tuffs, which are already rather thin here and due to lithological similarity, difficult to distinguish from the Middle and Upper Palembang beds.

The source of the Neogene sediments of eastern Sumatra, which according to Boissevain (1947) are more than 5,000 m thick in the deepest parts of the geosyncline, is the old Sundaland in the N, as appears from the homogeneous mineral association occurring until the end of the deposition of the Late-Oliocene to Early-Pleistocene Upper Palembang beds. The up-arching and volcanism of the Barisan Mountains is of a later date, having started during the Upper Palembang period. Folding of the geosyncline began after the deposition of the Upper Palembang beds, and NW/SE axes were formed. Intense base-levelling has since occurred, and it may be assumed that generally 1,000 m, and at the main anticlines even 3,000–5,000 m, of sediments were removed. The area was partly buried by a volcanic tuff cover, which at present is distinctly tilted.

The folding has, according to Boissevain, continued since then, as is shown by the fact that the anticlines, though not composed of more resistant materials, are topographically higher, and furthermore by the fact that subsequent rivers preferably follow the northern edges of the synclines, while drowned valleys and lakes or swamps occur in these zones. The present author's observations seem to confirm this. It will be evident from the above description that the area under discussion is not a peneplain in the true sense of the word but rather the result of a dynamic equilibrium of uplift and denudation. Smit Sibinga's (1951) view that it should be considered a piedmont plain is also justified, particularly if the important fill of the synclines is taken into consideration.

It appears to the present author that the important denudation following the orogenesis offers a satisfactory explanation for the coincidence of the northern limit of the Lampung tuff sheet with the limit of the block-faulted areas in the S. The cover has been removed in the folded, unstable areas further to the NW, where younger land forms prevail. Musper (1933) holds a rather different view, since he claims that the peneplain is preserved because it was protected by the volcanic tuff cover, which he believes to have greater resistance to erosion. The present author does not share this opinion.

The isolated outcrops of the Sundaland emerging from the eastern Sumatran lowlands have already been briefly mentioned. They occur near the mouth of the Kampar River, in the Batanghari plain, and at the eastern end of the Palembang anticline near Bukitbatu. Since the basement rocks in the surroundings must lie considerably lower, these occurrences are indicated on the geomorphological map as residual hills in analogy with the numerous residual hills rising from the peneplain of the islands E of southern Sumatra.

5. *Drainage pattern of the Palembang Lowlands* (Fig. 17)

Three main anticlinoria can be distinguished in southern and central Palembang, i.e. those of Muaraenim-Baturadja, Pendopo-Prabamulik, and Palembang. Further to the NW these three merge and become one broad folded zone. The ridges forming the first-mentioned anticlinorium rise considerably above the level of the peneplain and are therefore indicated as

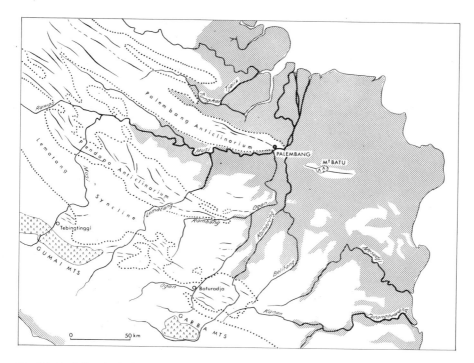

Fig. 17. Anticlinoria, drainage pattern, and alluvial plains of Palembang.

folded mountains on the geomorphological map. The Lahat-Muararaenim folds may be considered as the continuation of the Gumai Mountains discussed earlier, though they consist of younger sediments. E of Lahat they are pierced by the conspicuous Serilo and Asam andesitic plugs.

Terraces up to 40 m high are found along the rivers draining these areas. The stereogram of Photo 14 shows the low terrace (6 m) of the Lematang River at Lahat. The same terrace was also observed by the author further downstream, as well as a distinct 12 m terrace further upstream. The highest terraces are much more difficult to trace morphologically.

The eastward-flowing Kanan-Tulangbawang River collects the consequent rivers coming from the Barisan Mountains, as mentioned above (p 51), whereas on its left bank it is accompanied by a low divide and short tributaries. This divide is approximately in line with the northernmost anticline of the Baturadja anticlinorium, which plunges toward the E (see Van Tuyn, 1937). It is therefore possible that the present course of the Kanan River is determined by a weak fold axis. Further downstream, in any case, it flows in a wide syncline, the continuation of the broad synclinorium between the Baturadja and Pendopo anticlinoria.

A major change in the course of the Komering River after the Ranau eruption, is indicated by the distribution of the Ranau tuffs (ignimbrites)

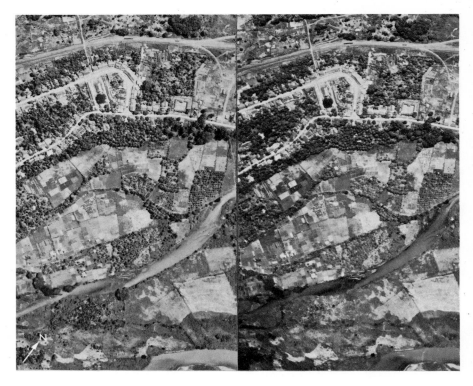

Photo 14. Stereopair of the Lematang terraces near Lahat in southern Sumatra. Scale about 1 : 10,000.

transported along the valley of this river. These deposits do not occur only along the present Komering course, but also at a somewhat higher level, in a belt in the Belitung resettlement area further to the E (Van Tuyn, 1937). The Lampoing River, located in the downstream extension of this ignimbrite belt, most likely follows the former course of the Komering (see geomorphological map). The Pisang River, which enters the ignimbrites of the Belitung plain, has been captured by the Kanan-Tulangbawang river system, which flows at a considerably lower level than the Komering River (Van Tuyn, 1937).

The country between the Kanan-Tulangbawang River in the S and the Mesudji River in the N has an elevation of only 10–20 m. The small rivers draining these '*talangs*'* are incised approximately 10–15 m in their upper courses, but frequently loose water downstream as a result of infiltration. The Mesudji River itself, after first running in a NE direction, arrives at a wide alluvial valley extending from the lower Komering to the ESE (the connection with the Komering was established long ago by excavation), and

* *talang:* local name for the low plateaus and ridges between the alluvial plains in the Palembang area.

then follows this valley to the sea. Although this valley runs nearly parallel with the main fold direction, it cannot be said whether it is structurally controlled. The lower course of the Mesudji River coincides with an extremely weak anticlinal structure (Van Tuyn, 1934b) in which, however, some faults occur.

The *talang* area N of this river was once probably an island, as indicated by the former shorelines (beach ridges, sand bars) bordering the Mesudji 'island' (61/63 x 13) on either side and reaching a height of 10 or more m above mean sea-level. Finds of old beads (*batu manik*) point to early settlements in this area. More recent beach ridges also occur, closer to the coast and at lower heights. The material of these sandy ridges was certainly derived from the *talangs*. The curved beach ridges to the E of the former island offer evidence that delta formation played no role in the formation of Cape Mendjangan; they indicate only a higher zone around which the beach was formed. Tectonically, the island, together with the lower Mesudji River, is situated on the very weak anticlinal structure already mentioned, which may be the end of the Pendopo anticlinorium dying out over the stable Sunda basement.

A broad depression stretches N from the Pendopo anticlinorium and can be traced eastward to the lowlands N of the 'island' of Mesudji. The greater part of this synclinorium is drained by the Musi River, but where this river turns northward it is continued by a zone with swamps and lakes and by the eastward-running Lumpur River.

This zone is bounded on the N by the Palembang anticlinorium, the *talang* area forming an elongated Mt Seguntang 'peninsula' in the alluvial plain with the town of Palembang at its extreme end. The extension of this ridge is found to the E of the Musi River where the Mt Batu granitic outcrop occurs. This represents an isolated residual hill of the Sunda peneplain (p 7), a comparatively stable zone underlying this area. The fact that this structural feature is also reflected in the alluvial plain evidently implies that the tectonic movements continue to the present day, a view which is consistent with the observations made by Boissevain, Musper, and other investigators elsewhere in southeastern Sumatra. The occurrence of the Tungkal and Lalang rivers in the alluvial plain NE of Palembang and occupying synclinal zones (Smit Sibinga, 1951) point in the same direction.

The Musi River, joined by the Ogan and Komering rivers, breaks through the Palembang anticlinorium where an axial depression occurs near the town of Palembang. A similar depression is located further S in the Pendopo anticlinorium, and the Komering and Ogan rivers thus flow in broad valleys filled with alluvium and maintain their consequent character over the entire distance from the Barisan Mountains to the coast.

6. *Ancient coastlines*

An interesting attempt at a reconstruction of ancient coastlines in southeastern Sumatra, based on historical data from such sources as Chinese chronicles, was made by Obdeyn (1941/43). Although there may be some doubt regarding some of his conclusions on geological grounds, it is evident that only a narrow alluvial plain existed to the E of Palembang in the days of the Srivijaya kingdom. Obdeyn draws part of his coastlines in areas where Neogene rocks outcrop, which is obviously incorrect. He estimated the period elapsed since the beginning of the formation of the alluvial plain at approximately 2,000 years. This figure is based on the linear accretion at the Batang Hari mouth and on historical records of questionable reliability (see also Tjia et al., 1968). The age of the alluvial plain around Djakarta has been estimated by the present author at 5,000 years, on the basis of the rate of areal accretion. Although this figure may be too high, the 2,000 years mentioned for southern Sumatra by Obdeyn seem an absolute minimum. These figures in any case leave no room for the deposition of earlier Holocene sediments, which is a strong argument in favour of the present author's view that the width of the alluvial plain was determined by sea-level changes and crustal movements rather than by the rate of sedimentation of the rivers.

The natural levees ('*renahs*') of the present and former river courses are often the only dry parts of the otherwise swampy plain, and so settlements are located on them, although houses built on rafts are also a common feature. Swamp rice is grown on the black soils of the back-swamps, the drainage of which could still be considerably improved. Stagnant water is found in the deepest parts, which are surrounded by extensive swamp forests. A narrow belt of dry-land forest marks the transition to the above-mentioned higher and reddish *talang* areas, where natural vegetation is found only in the valleys and ravines.

7. *The northwestern parts of the peneplain*

The rivers draining the peneplain area further to the N are also distinctly influenced by the geological trend-lines. The Pendopo anticlinorium has an important effect in this respect: the Rambang River flows to the E and then turns around the plunging eastern end; and the Lematang River crosses it at an axial depression (55.5 x 13). Further to the W, the Musi River crosses various axes of the anticlinorium of Pendopo Mountain at axial depressions and is then joined by the Hari Leko River, which runs along the southern flank of the Palembang anticlinorium. Still further to the NW, the extensive swampy area along the Rawas and Rupit rivers before they cross the western end of the Pendopo anticlinorium, is also clearly structurally controlled.

The value of these rivers for irrigation purposes depends largely on the amount and quality of their suspension load. For instance, the Klingi resettlement area, located near the borderline where the Kaba fluvio-volcanic fans reach the lowland, produces excellent dry crops, whereas the rice harvest is only moderate, due to the rather poor Klingi water. The reverse situation occurs in the Belitang area, where only moderately fertile soils are favoured by the excellent Komering irrigation water. The Sekampong River in the Lampung Districts also carries a sizeable and fertile load, from which the Metro resettlement area benefits.

VI Central Sumatra

a. Introduction

The main structural geomorphological zones distinguished in southern Sumatra can also be traced in the central portion of the island. The geomorphology here is in general somewhat more complicated, however, particularly W of the Median Graben. An equivalent of the gently oceanward-tilted Bengkulu Block cannot be distinguished N of Padang. It is replaced by a more complicated arrangement characterized by alternating high blocks and alluvial plains indicating submerged blocks, the abrupt transition producing steep mountain slopes and a sudden decrease in the gradient of many rivers upon entering the alluvial plain at the foot of fault scarps. Where volcanic or other sources provide these rivers with a sufficient load of coarse-grained material, large alluvial fans are formed.

The Median Graben is usually well developed in central Sumatra and can be traced from southern Kerintji, via the Solok plain and Lake Singkarak, to the 'Padang Highlands' of Bukit Tinggi and further N to the graben of the Sumpur and Angkola rivers (Figs. 18 and 24).

High stratovolcanoes occur in the Barisan Mountains, although they are distinctly less numerous than in southern Sumatra. Their products cover considerable parts of the block mountains, and large fluvio-volcanic fans reach the peneplain E of the Barisan range through a number of gaps in the complex mountains. These gaps are comparable to those of Sugiwaras and Muarabeliti mentioned for southern Sumatra. Further to the NW, the Barisan becomes narrower and volcanism more localized. The peneplain of the eastern part of the island is the continuation of the one in the Palembang/Djambi area of southern Sumatra. A somewhat lower zone, the so-called Sub-Barisan Depression, often occurs immediately to the E of the Barisan range. Pre-Tertiary rocks crop out in the highest upfolded parts of the geosyncline, predominantly in the Tigapuluh Mountains. Geomorphological features of special interest in this zone are the low cuestas or hogbacks formed locally in more resistant Tertiary beds and the structurally controlled drainage system. The alluvial plain extending to the E is broad and often swampy. Extensive peat deposits occur in the back swamps.

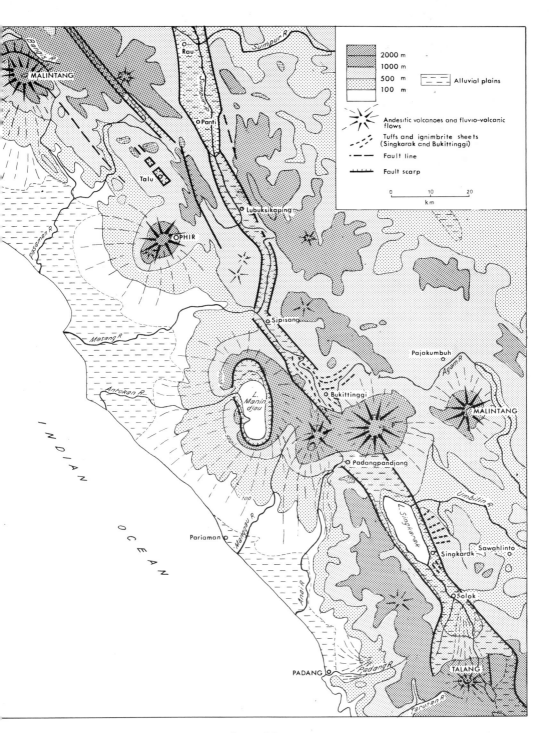

Fig. 18. Sketch map of the western part of central Sumatra.

b. The Block Mountains to the West of the Median Graben

1. The continuation of the Bengkulu Block
The southernmost part of this zone in central Sumatra is still largely an oceanward-tilted block drained by a number of consequent rivers. One of these, the Dikit River, is noteworthy because its upper reaches drain in the Median Graben and break through the western fault-scarp, which is considerably lower here than the eastern one; some minor extinct volcanic cones occur at this locality.

At two places along the Bengkulu Block, large fans border on the coastal plain. The larger and probably older one, near the Silaut River, is certainly fluvio-volcanic and was once formed by now extinct volcanoes located upstream. It is unlikely that the Lake Kerintji eruption (Verstappen, 1955) contributed to its formation. The material of the other fan, at the lower end of the Batang River, derived from the stratovolcano, Mt Kerintji, lying E of the Median Graben, and thus passed the graben to reach the western coastal plain.

This section of the Bengkulu Block (and the same applies to the section more to the NW toward Padang) contains various longitudinal ridges running parallel to the coast. Since several rivers are deflected by these ridges before breaking through them (e.g. the Batang River), longitudinal faults may have influenced the drainage pattern here.

2. Indrapura plain
A remarkable feature in this area is the wide triangular coastal plain which broadens to the NW, where it terminates rather abruptly. Since it is rather swampy, settlements are preferably located where rivers emerging from the mountains have formed fans. This plain is a larger replica of the alluvial plain SE of Bengkulu in southern Sumatra (p 28). Its position must be determined by some underlying structure, because there is no other reason why the rivers here should have built a much wider alluvial plain than the ones formed further to the N and S. A pseudo barrier reef extending in a northwestern direction forms the continuation of the seaward limit of the alluvial plain, which can thus be considered as a fill of the southernmost extremity of the lagoon. This lagoon was apparently shallower or was filled in earlier here than further to the NW, which could mean that the area is underlain by a submerged block dipping in this direction.

The swampy plain is drained to the NW, parallel to the coast, by the Indrapura River, which collects the consequent rivers coming from the Barisan range. This direction is probably determined by beach ridges parallel to the coast.

3. Padang Section

The alluvial plain further to the NW is narrow and finally tapers off. From this point onward the coastal area consists of steep Old Andesitic hills, which make the rocky coast extremely irregular, with numerous capes, little bays, and islets. This mountainous area is bordered to the N by a steep escarpment marking the edge of the tilted block, the mountains further on being much further away from the sea. The escarpment is largely influenced by faults, and a number of these are clearly reflected in the morphology and drainage (Fig. 18).

This escarpment at the same time marks the beginning of the coastal plain in which the city of Padang is situated and which continues to the point where the Manidjau tuff flows reach the coast. Numerous beach ridges lying approximately parallel to the coast can be traced in this plain.

Fig. 19. Geomorphological sketch of the area between Padang and the local cement factory; scale 1 : 125,000. Numerous faults (FF) in the mountainous areas strongly influence the drainage pattern. Huge volcanic mudflows (P1 to P4) of unknown source occur in the valleys; note the radial drainage of the fan-shaped flows. No irrigated rice fields are found in P1, which is undoubtedly the oldest of the flows. Few irrigated fields occur in the P2 flows, which are probably slightly younger, whereas the still younger P3 and P4 flows are widely irrigated. Note the short narrow gorge of the Padang River where it breaks through a fault scarp near the village of Lubuk. (cf. Photo 15). *Key: 1. Surrounding mountains; 2. Fluvio-volcanic fans P1; 3. P1a; 4. P2; 5. P3; 6. P4; 7. Beach ridges; O.R.* = Old river course; F = Faults; A = Monkey Hill; B = Telukajur tombolo.

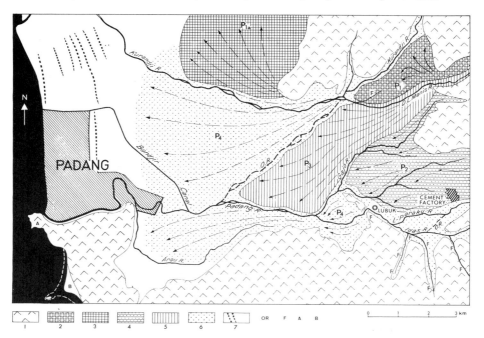

4. *Fluvio-volcanic fans of the Padang plain*

A complex of huge alluvial fans enters the plain directly to the N of this escarpment near Padang, as indicated in Fig. 19. These features are almost certainly fluvio-volcanic in origin, but their source, remarkably enough, is unknown. Remnants of old fans in the northern part of this multi-cycle fan system definitely represent the oldest part (P 1–P 1A). The pattern of shallowly-incised radial dry valleys in these oldest fans, which are traceable from the air, indicates that the material originated from the upstream parts of the Danau-Limau-Manis valley. It is noteworthy that the Kurandji-Danau-Manis river system formerly drained to the SW and debouched into the Padang River (OR in Fig. 19), but has since been diverted WNW.

Photo 15. Stereogram of the mudflows in the Arau valley between Padang and the local cement factory, to scale 1 : 60.000. Compare with Fig. 19. Note the radial drainage and the differences in appearance of the various flows of different age (P1–P4). The straight and abrupt transition from the mountains to the valley bottom near the village of Lubuk (L) suggests a fault scarp where the Padang River breaks through in a short narrow gorge.

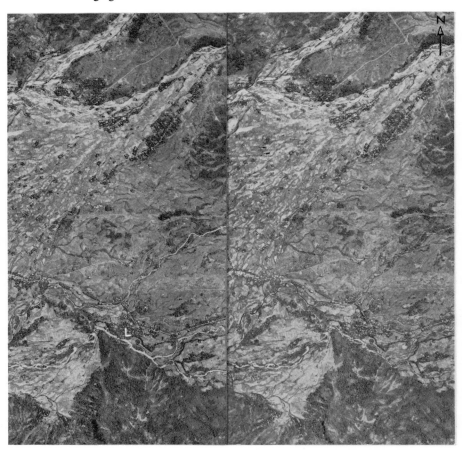

An important old fan relic (P 2) located along the headwaters of the Padang River must be of the same age or slightly younger than the relic P 1, but the dry valleys occurring in it indicate that the material came from the Padang River in the E.

A distinctly younger and lower relic (P 3) can be seen between P 1 and P 2 along the former dry Kurandji course (OR). Its radial drainage pattern again points to the upper Danau-Limau-Manis valley as a source area.

The youngest and lowest parts (P 4) of the fan system can be divided into two parts, according to their source area. The parts to the N of the Padang River show a radial drainage having its apex in the Kurandji-Danau-Manis river system. The fact that the Kurandji River was diverted from the SW to the NNW after the formation of P 4, suggests that the material of this fan also came from the Danau-Limau-Manis valley. The material of the P 4 areas in the S certainly originated from the upstream parts of the Padang river system.

The stereogram in Photo 15 shows part of this remarkable area. It is interesting to note how two headwaters of the Padang River jointly break through a NW-striking fault scarp near the village of Lubuk and form a narrow gorge in the straight ridge occurring there.

The upstream parts of the fan relics are considerably more deeply incised by the present rivers than their downstream parts. This phenomenon is easily accounted for by the concave profile of the present rivers, whose lower courses have a gentler gradient than the somewhat steeper slopes characteristic of the mudflow surfaces.

Further to the NW the mountains recede, the plain broadens, and we arrive at another fan extending into the coastal plain. This fan originated from the Tandikat-Singgalang twin volcano situated near the Median Graben. At this place the Anai River breaks through the fault scarp and the western block mountains in a deep gorge through which the railway and the road connecting Padang with the Highlands pass. In the gorge, Tertiary rocks are still exposed. The fluvio-volcanic flows from Mt Tandikat must have pushed the river against the block mountains, and even in the plain the river continues to run at the foot of these mountains before emptying into the Indian Ocean.

The drainage changes in the upper course of the Anai River will be discussed in connection with the Median Graben.

5. *Lake Manindjau cauldron*

The Manindjau cauldron (34.5 x 12) carrying the lake of the same name and measuring approximately 8 x 16½ km, is located NW of this group of stratovolcanoes. This depression was the scene of huge volcanic eruptions of the type known from the much larger Toba cauldron. A large area is covered by the fluvio-volcanic material produced by these eruptions, particularly to the NW, W, and SW of the cauldron. A radial drainage pattern, mostly shallowly incised, is characteristic of this area. The Manindjau fluvio-volcanics cause a gradual narrowing of the alluvial plain until they reach the coast N of Pariaman. A description of the cauldron is given by Kemmerling (1920).

The fluvio-volcanics cover a considerably narrower zone in the NE and E, where their outflow was hampered by older mountain ridges. Older rocks frequently emerge from the tuff plateau, which is incised here by numerous rivers with deep valleys bordered by vertical sides.

The rim of the Manindjau cauldron reaches heights of 1,200–1,400 m above sea-level; according to a survey by Verbeek (1883) the lake level is at 459 m and the greatest depth 157 m. The steep slopes bordering the flat-bottomed lake are thus hundreds of metres high, the combined result of faulting and volcanic explosions. The slopes are particularly steep in the S, where a peninsula marks the edge of the southernmost and youngest eruption centre. The slopes surrounding the northern portion of the lake are somewhat less steep, and a lakeward-sloping plain with irrigated rice fields is found here between the escarpment and the lake. It seems that the lake level was formerly somewhat higher, although no relics of lake terraces could be traced by the author, possibly because of more recent mass movements along the steep slopes, at the foot of which steep scree fans and minor deltas occur. The only indication of a former higher lake level occurs in the NW, where the Antokan River drains the lake to the Indian Ocean. This uncertain level lies at 570 m above sea-level.

6. *Ophir Section and adjacent lowlands*

The last section of the western zone falling within central Sumatra is dominated by the lofty Ophir (Talakmau) volcano (2,912 m) rising directly from the lowlands and forming an impressive and slender cone. This is actually a complex volcano, because two lower extinct cones are situated on a line running to the SW. The numerous crater pits, etc., of the main cone also

show the same alignment. No reliable evidence of recent activity is known (Neumann von Padang, 1940).

The Ophir volcanic products extend far into the wide coastal plain, which penetrates inland on either side. Both parts are swampy and have beach ridges along the coast. A series of older beach ridges can be seen near the landward rim of the swampy plain behind Airbangis, which indicates that the coastline was formerly much more irregular. The extremely swampy nature of this alluvial plain could be due to recent relative subsidence, which might also explain the present recession of the coast near Airbangis.

Various isolated hills emerging from the coastal plain consist mostly of Permo-carboniferous rocks, and in the W also of granites and andesites, e.g. near Airbangis (Photo 16).

The western block mountains show a pronounced difference from the previously mentioned areas further to the SE: beds of Permo-carboniferous limestones lying parallel to the Sumatra direction form high ridges which rise from the peneplain forming the surface of the block mountains and which are traversed by the rivers. Moreover, the mountains in this northwestern section are cut by longitudinal faults and graben structures giving rise to a structurally controlled drainage pattern. One of these grabens is located a few kilometres W of the Median Graben, which it parallels. Another graben, running in the same direction, contains the depressions of Bandjar (Fig. 20) and Talu. A tuff-covered ridge separates these two depressions, which contain important tuff deposits and were probably the site of fissure eruptions. A hot spring occurs in the Talu depression. Several faults in the Sumatra direction can be traced from the air in this zone.

Photo 16. View of the area SE of Airbangis as seen from Mt Marando. The vast alluvial plain marks a submerged block of the Barisan Mountains from which the andesitic Mt Sitjangang (centre) and several Permo-carboniferous hills (e.g. Mt Tulas) rise.

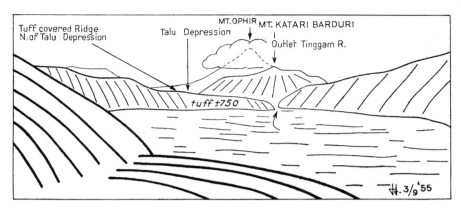

Fig. 20. Field sketch of the Bandjar volcano-tectonic depression, looking southward. This structure is located NW of the larger Talu depression, from which it is separated by a tuff-covered ridge. The Ophir volcano appears in the distance.

c. The Median Graben

1. *The parts southeast of Lake Kerintji*

This zone is rather poorly defined in the southernmost part of central Sumatra. The narrow graben located further to the S (p 39) and drained toward the Indian Ocean by the Seblat and Dikit rivers can hardly be traced SE of Lake Kerintji. The western fault scarp here is rather well developed, although it is partially buried by volcanoes. The eastern fault scarp can only be indicated with difficulty, however. Some ridges occur several kilometres to the E of the western fault scarp, but it is more likely that the eastern fault scarp must be sought in the higher grounds occurring to the N of the Merangin River, which drains Lake Kerintji.

This part of the graben is widely covered by volcanic ashes deposited by the surrounding volcanoes. A rather small volcano, Mt Tua (44.5 x 9.5), is located in the graben and is noteworthy because of its well-preserved cone and because of the surrounding explosion craters holding lakes. Tuff-flows extend eastward and reach the incoherent plain of Lologadang in South Kerintji. This plain is drained by the Lintang River through a narrow outlet only, and increased soil erosion upstream has led to deteriorating drainage conditions in this area. A terrace occurs at 960 m above sea-level, which indicates that the river was blocked and a lake ponded up here immediately after the Lake Kerintji (Gadang) volcanic eruption to be discussed below. There are also some indications that part of the area once drained toward the large Kerintji plain further N.

2. Kerintji intramontane plain

The Kerintji intramontane plain is a wedge-shaped complex belonging to the Median Graben and lying at an altitude of about 800 m (Fig. 21). It is bordered by steep fault scarps with a height of approximately 400 m and converging towards the NW, where the narrowest part of the graben is located. The plain thus becomes broader and also lower toward the SE and reaches a maximum width of about 9 km at the extreme southeastern end, where the volcanic Kerintji (Gadang) Lake, with a maximum depth of 110 m, is located.

A number of secondary faults are also geomorphologically distinct and affect, for example, the drainage pattern W of the graben, as shown by the straight sections in the Sumatra direction of various rivers (Verstappen, 1955). Some straight river courses to the E of the plain represent a comparable situation.

Two splinters occur within the northern part of the graben and contribute to its narrowing toward the NW (Fig. 21). One of these is marked by the occurrence of hot springs. A row of hot springs also occurs along the main eastern

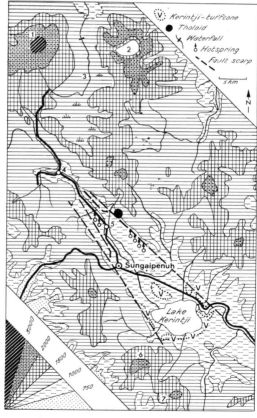

Fig. 21. Contour map of the wedge-shaped Kerintji Graben and surroundings. Heights are given in metres.

Key: 1. Kerintji volcano
2. Tudjuh volcano with crater lake
3. Bento marshes
4. Siulakderas
5. Semurup
6. Raja volcano
7. Kunjit volcano

Photo 17. Rhyolitic bocca, forming steep-sided hill at the foot of the eastern fault scarp of the Kerintji graben near Pendung.

Fig. 22. The surroundings of Lake Kerintji, scale 1 : 100,000
 Key: 1. Kerintji tuffs
  2. Mountainous areas
  3. Slope of Bamban volcano
  4. Scarps
 Depth contours (10-m interval) from a lake survey by the author.

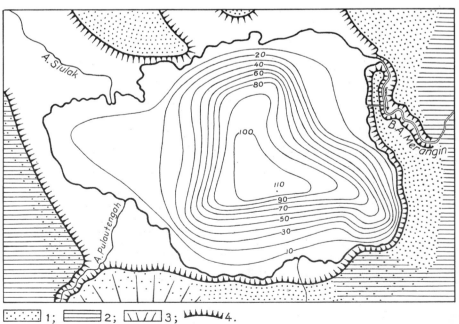

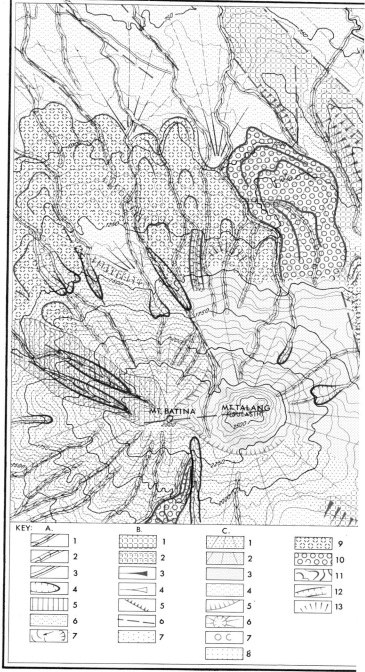

Fig. 25. Detailed geomorphological map (1 : 75.000) of the Mt Talang volca di Baruh Lake, central Sumatra.

Key: *A. Forms of exogenous origin*
  1. V-Shaped ravines
  2. U-Shaped ravines
  3. Shallowly incised ravines
  4. Main barrancos
  5. Gullying influenced by faults
  6. Alluvial plains
  7. Landslides

B. Forms of structural origin
  1. Horst
  2. Graben
  3. Fault scarp
  4. Eroded fault scarp
  5. Minor fault scarp
  6. Fault line
  7. Fault scarp covered by slid

Photo 18. Tuff terraces around Lake Kerintji, as seen from a hill near the outlet of the lake. Kotapetai hill can be seen on the far side of the lake and the western fault scarp bordering the graben appears in the distance.

fault scarp. A rhyolitic bocca forming a steep-sided hill originated on the same fault (Photo 17).

A volcanic eruption must have occurred in the broadest southeastern part of the plain, the ash deposits of which can be observed around the present lake, as indicated in Fig. 22. Distinct terraces formed in these products (Photo 18), indicating that after the eruption the whole plain was flooded up to a heigt of about 90 m above the present lake level. Rapid incision in the ashes by the Merangin River, which drains the lake, soon caused a lowering of the level, after which the plain ran dry again. The terraces can also be traced along this river (Fig. 23).

Further subsidence of the graben occasionally occurs during earthquakes,

Fig. 23. Terraces in Kerintji tuffs in the Merangin valley; looking downstream from a point at about 1 km below the outlet of the lake. The road from Sungei Penuh to Banko is situated on the left.

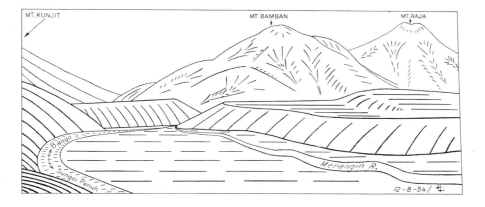

as for example in 1909, and the obstruction of the Merangin River then results in a rise in the level of the lake and drowning of the rice fields bordering it. The hydrological situation is thus critical, and man-made erosion of the steep fault scarps bordering the plain has a very harmful effect on conditions in the plain.

The graben finally terminates at the North Kerintji granite and quartzdiorite massif, where the further continuation of the faults becomes rather obscure. The poor development of the northern part of the Kerintji graben may be related to the presence of this granite massif. Similar situations are known to the author from other parts of the Barisan Mountains. The small Siulakderas basin (42 x 10.3) is formed in the southern part of this massif partly by selective erosion of softer shaly and gneissic streaks in the quartzdiorites and also by tectonic movements, as suggested by the peculiar drainage of the area. Various terraces above the plain near Siulakderas, which reach heights of 7 m and 15–20 m above the river, are most probably drift terraces related to an andesitic rock sill which, further downstream, separates this basin from the Kerintji plain.

The western fault scarp can be traced further to the NW, but the eastern fault scarp disappears under the young tuffs of Mt Kerintji, only to reappear at the other side of this volcano. To the NW of the sources of the Siulak River, which drains the graben toward Lake Kerintji, are the upper reaches (with two small lakes and a tiny plain) of the Batang River, which breaks through the western block mountains and flows into the Indian Ocean. Where it reaches the plain, the alluvial fan mentioned in section 6b, has accumulated, having derived most of its material from the Mt Kerintji area.

3. *Batang Hari Section*

The graben section N of Mt Kerintji is drained by various headwaters of the Batang Hari (Batang = large river), which meet in the eastern lowlands to form this largest river of Sumatra.

Although the graben has a regular NE–SW slope toward the deepest point at Muaralabuh (Fig. 24), there are no fewer than four breaches in the eastern fault scarp through which rivers escape from the graben in an eastern direction.

At the deepest point at Muaralabuh, a river from the SW joins the Siliti River coming from the NE, which then breaks through the eastern scarp. Flowing parallel to the upper reaches of the Siliti River, however, there is a second river in the graben that breaks separately through the eastern scarp, whereas the Siliti continues its way to the SW. Still higher up, the Batang Hari itself flows through the graben, but this river breaks through the eastern scarp near the place where the Siliti takes its origin within the graben.

This peculiar situation, which deserves a more thorough investigation, is probably due to the fact that a number of parallel faults occur in this area instead of a well-developed graben. The complexity of this graben section

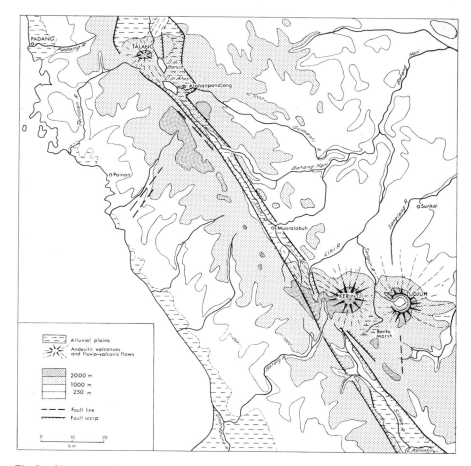

Fig. 24. Sketch map of the southwestern part of central Sumatra.

is also demonstrated by the narrow depression zone with irrigated rice fields situated approximately 1 km W of the Batang Hari.

4. *The Baruh and Atas lakes*

The Median Graben reaches its greatest height of about 1,500 m at the well-known lakes Danau di Atas and Danau di Baruh (38.5 x 11.5). Numerous faults, some of which acted as volcanic vents, border these volcano-tectonic lakes. It is noteworthy that the southernmost of these lakes, Danau di Atas, is now drained by the Gumanti River toward the E. Originally, however, it was drained toward the Median Graben in the SE by the Batang Hari. A shallow saddle in the tuffs surrounding the lake and lying at about 9 m above the present lake level, indicates this old outlet. The Gumanti River, having a steeper gradient, must have captured the lake, the level of which has since dropped 9 m, as shown by terraces formed on the lake shores,

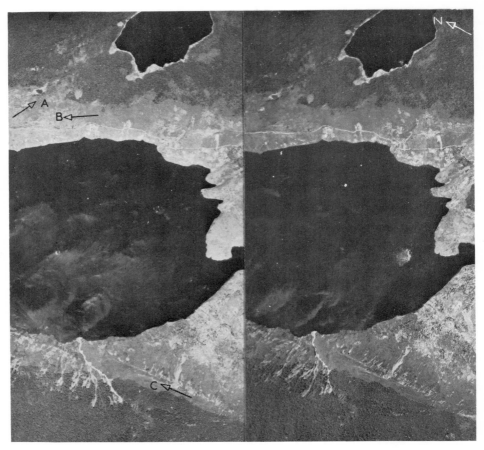

Photo 19. Stereogram of part of Lake Baruh and Lake Talang situated near the Talang volcano. Scale about 1 : 51,000. Note the scarplet (A) marking the fault running along the SW side of Lake Talang. Lake Baruh is bounded on either side by the faults of the Median Graben. Young scarplets having influenced the minor drainage elements can be seen at B and C.

particularly on the northern and northeastern sides. Several small rivers are diverted by the tuff deposits surrounding the lake and pass it at a short distance. A small lake basin, which has since run dry, is located in the tuff deposits between the two lakes.

Danau di Baruh is drained by a river flowing NW toward the next section of the Median Graben (Fig. 24). A terrace remnant indicates a former lake level 15 m higher than the present one. Whereas the shallow Danau di Atas is surrounded by tuff deposits reaching only a rather limited height, the tuff deposits of the considerably deeper Danau di Baruh are much more extensive. The tuff-covered terrain E of this lake gradually slopes down to a plain with irrigated rice fields, extending as far as Alahan Pandjang. This plain may well be a feature similar to but of greater age than the two lakes under discussion.

The fault scarp bordering Danau di Baruh on the E, indicated in Fig. 25, has been partially obliterated by a major landslide and is considerably eroded directly to the N, where the lake reaches its maximum width. It is, however, very well preserved and almost undissected further to the N, where it gradually decreases in height. Two scarplets occurring mid-slope indicate continuing fault movements in more recent times. Since these scarplets face the main fault scarp, it is evident that the graben side has been slightly uplifted with respect to the block bordering it to the E.

A comparable situation exists along the much lower fault scarp bordering the lake to the W. The same fault also borders Danau Di Atas, but on its eastern side. Here, the downthrow is thus the precise opposite of the situation described above. A young inverse fault scarp can be traced parallel to the western shore of Danau di Baruh, however.

A feature of particular interest here is a small lateral graben directed WNW and thus making an acute angle with the Median Graben. This graben contains the tectonic Lake Talang. The various features can be clearly seen in the stereogram of Photo 19 and in Photo 20, which gives a ground view of the western fault scarp of Danau di Baruh and the scarplet bordering the small Talang graben to the S.

The Median Graben NW of Danau di Baruh is rather complex, and several faults occur even in young volcanic and fluvio-volcanic material. A local horst runs from the northern lake shore in a northwesterly direction. The outcropping rocks are mainly lavas, partly covered by ashes.

E of the graben the tuff cover is much thicker and probably originates from the Baruh eruption(s). At its top is a hitherto unknown cinder cone.

Photo 20. Ground view of the area of Photo 19. Lake Baruh is seen at the right and centre; the Talang volcano appears in the distance to the left. Note the scarplet (A) bordering Lake Talang (not visible).

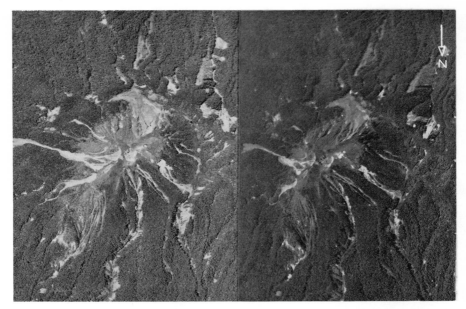

Photo 21. Stereogram of the top area of the Talang volcano (scale about 1 : 50,000). Note the SW–NE alignment of the recent crater pits of the young cone, which is surrounded on three sides by an older caldera. The latter influences the location of both the rivers draining the area and the volcanic ash and mudflows coming down from the young cone.

5. *Talang volcano*

The active Talang volcano (2,597 m) is situated not far from the western fault scarp of Lake Baruh; the scarp disappears beneath the volcanic mantle and cannot be traced to the N of it in the Solok plain. Another fault scarp here forms the western scarp of the Median Graben. Since it passes W of the volcano, the latter is actually situated within the graben zone, which must be rather complex here.

The volcano actually forms a twin with the extinct Mt Batina located further W along an E–W-striking fault playing an important role in the development of the volcano. The active Talang cone is situated in a caldera of which only the western and southern rim are left. Since the cone rose up in this caldera after the partial collapse of the latter (Photo 21), the young Talang products can only spread freely toward the N and E. A number of solfataras and small crater pits are found in an E–W-running depression at approximately 2,400 m altitude, demonstrating the lasting influence on the volcanic activity of the previously-mentioned fault (Fig. 25). Several recent and unvegetated lahar flows radiate from the top area. A radial pattern of narrow ravines is characteristic for both the younger and older parts of the cone. The dissection is deepest in the ashes at their upstream end. Incision is particularly deep in ravines located along one of the radial faults occurring in the volcanic body. A segment of the western slopes, for instance, has been considerably lowered by the more severe erosion along faults.

Lava-flows are a dominant feature of the lower slopes, particularly in the N. Their sources, which lie almost exclusively in the lower slopes, are covered by more recent ashes. Flow structures and pressure ridges are easily visible in many localities and 'grooves' representing collapsed lava tunnels are common features. Three flows, directed N, S and ESE, are somewhat more distinct and therefore probably younger than the others. The last-mentioned flow enters the northern end of Danau di Baruh. Fluvio-volcanic fans occur predominently beyond the lava belt, and are indicated in Fig. 25, which is based on photo-interpretation and a partial field check by the author.

The nature of the fault scarp passing W of the Talang volcano is complex and still poses many problems for future research. One branch runs in line with the southwestern shore of Danau di Atas and is followed by the upper course of the Tarusan River (Fig. 24), which then breaks through it to flow toward the Indian Ocean. The scarp then turns to the NNE and is followed by a river directed towards the Solok plain; this river has evidently been pushed against the escarpment by the volcanic flows from the Talang volcano. A zone of numerous steep hills rising from the rice fields to the S of Solok is a remarkable phenomenon that probably represents a major landslide related to the collapsed northeastern flank of the old Talang caldera.

6. *Solok-Singkarak Section*

The Solok plain is at present drained to the NW, toward Lake Singkarak, which in turn is drained eastward by the Umbilin River. There is, however, as observed by Van Valkenburg (1921), a dry river valley to the E of the town of Solok. Evidently, the Solok plain once drained directly eastward via this now deserted valley, which reaches the Umbilin River many kilometres further downstream. The height of the dry valley is about 410 m, equalling the higher terraces at the Umbilin outlet of Lake Singkarak (Photo 22). The terraces observed here by the author reach heights of 5, 12, 32, and 57 m above the river, which here lies at 355 m above sea-level. The two above-

Photo 22. Terraces along the Umbilin River near the outlet of Lake Singkarak, which is just visible in the distance.

Photo 23. Lake Singkarak as seen from its northern end. The western fault scarp and Talang volcano are in the background; the area in the foreground forms part of the tuff deposits surrounding this end of the lake.

mentioned passages obviously were in use simultaneously. The present northwestward drainage of the Solok Plain toward Lake Singkarak (Photo 23) can only be explained by assuming a recent subsidence of the graben section between Solok and the lake. Several geomorphological features offer evidence of very recent faulting movements in this part of the Median Graben, shown by the geomorphological sketch in Fig. 26. These changes are essentially due to

Fig. 26. Geomorphological sketch of the Median Graben between Solok and Singkarak, central Sumatra, to scale 1 : 125,000.
    Key: 1. Mountainous areas beyond the graben
         2. Tuff deposits resulting from a fissure eruption in the graben and probably covering a less deeply-subsided graben section
         3. Alluvial plain within the graben
         4. Scarplet resulting from renewed tectonic activity and marked in the field by the sites of several villages
         5. Young fault lines

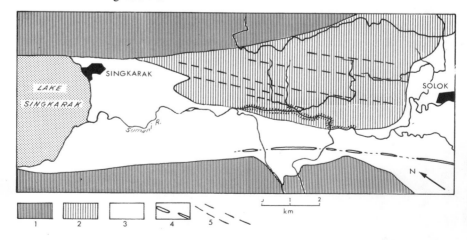

Photo 24. Alluvial plain SE of Lake Singkarak with tuff deposits in the Median Graben between Singkarak and Solok (see also Fig. 29), looking SE. The fault scarp bordering the graben to the W and the Talang volcano are visible in the distance on the right.

tectonic movements, though some volcanic activity was associated with the faulting (Verstappen, 1961). Young whitish volcanic tuffs cover parts of the graben N of Solok (Photo 24), as can be seen from Fig. 26. They are not related to any volcanic cone, but originated from a fissure eruption at the eastern side of the graben accompanying the recent collapse of this graben section. The tectonic movements must have continued after the fissure eruption, because a number of well-preserved scarplets with an influence on the drainage can be seen in the tuffs. The Sumani River breaks through the tuffs located in the central and eastern parts of the graben, although a course through the flat and lower westerly plain would have encountered less impedance. This points to an even more prolonged subsidence of the western side of this graben section, a view which is also supported by the occurrence of a young and straight scarplet in the rice fields in front of the western fault scarp. The scarplet can be traced from a point S of Solok to near Lake Singkarak, and a number of villages have been built on top of it.

Photo 25. Ignimbrite/tuff cover, sloping down to the E and originating from a fissure eruption along the fault scarp bordering Lake Singkarak on the E. The lake is situated beyond the right edge of the photograph, which was taken looking SE from a hill at Padangdataran near the outlet of the lake.

Photo 26. The northern end of Lake Singkarak with the tuff deposits occurring there (see Photo 23). The western fault scarp and the Anai gorge (right) are in the distance.

The fault bordering Lake Singkarak to the E also acted as a volcanic fissure. The tuffs/ignimbrites were deposited to the E of the fault scarp and form a surface sloping gently eastward (Photo 25). Another but smaller flow is found further S. A third tuff deposit related to the Singkarak eruption is a tuff cone in the graben enclosing the northern side of the lake (Photo 26).

7. *Bukittinggi ignimbrite plateau*
N of Lake Singkarak the graben is buried beneath the slopes of the Marapi (2,891 m) and Singgalang (2,877 m) volcanoes, whose lower slopes intersect and thus form a saddle acting as a divide.

The rivers coming down from the southern slopes of the Singgalang volcano must once have flowed to Lake Singkarak but have been captured in comparatively recent times by the Anai River, and since then have drained toward the Indian Ocean in the W (Fig. 27). The pirate Anai River is deeply incised in a narrow gorge. The capture occurred near the town of Padangpandjang (36 x 12), and the present, almost imperceptible divide between the Anai branches and the beheaded Sumpur River, which still flows toward the lake, lies at about 770 m above sea-level.

The western fault scarp is well developed in this section, whereas the eastern

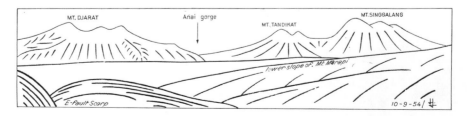

Fig. 27. Field sketch showing part of the Median Graben NW of Lake Singkarak, as seen from the pass between Padangpandjang and Batusangkar on the eastern fault scarp. The gorge of the Anai River, which drains this graben section towards the Indian Ocean, is seen in the distance; the Singgalang volcano appears on the far right.

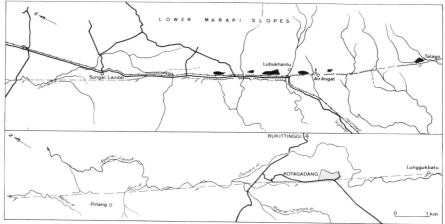

Fig. 28. Recent fault scarp (F) in the lower slopes of the Marapi volcano in central Sumatra, passing along Kotagadang near Bukittinggi and continuing NW toward the Matur valley. Note the scarplet, the deranged drainage pattern, the lakes and ponds, and the hot spring near Airangat. The railroad and the highway were constructed exactly along the fault between the villages Lubukhantu and Sungailandai.

one is rather low and disappears underneath the Marapi volcanic slopes, only to reappear to the NW of the Bukittinggi ignimbrite plateau.

A geomorphological feature of particular interest is an extremely young fault scarp many kilometres long occurring in the loose ashes of the western Marapi foot (Fig. 28). For some distance the road and the railroad have been built exactly over this young fault. Although this situation is readily explained by the fact that fault-line erosion in mountainous areas sometimes creates the easiest passages, it could be dangerous in the event of an earthquake. Similar situations occur elsewhere in the Barisan Mountains. The road and railroad diverge from the fault further to the NW, where it can be traced along various river courses, among them the famous Ngarai Sianok River valley or 'Water Buffalo Canyon' mentioned below.

Photo 27. The steep-sided Ngarai Sianok valley ('Water-Buffalo Canyon') formed in the Padang Highland ignimbrites as seen from Bukittinggi, looking W. The Singgalang volcano appears in the distance on the right, and the slopes of the Marapi volcano are just visible at the extreme left.

It is noteworthy that the western limb of the fault has been raised in relation to the Marapi volcano, which resulted in the formation of small ponds and swamps and in diversion of the radial rivers draining the Marapi cone, but that further to the N and S the eastern limb is higher, which caused the diversion of rivers draining the Singgalang volcano and a small escarpment bordering rice fields near Kotagadang. Many of these details are even distinctly visible on the 1 : 40,000 topographic map of central Sumatra.

The Bukittinggi intramontane plain located N of Mt Marapi, at an altitude of about 900 m, is covered by an ignimbrite sheet which is older than the andesitic stratovolcanoes. The ignimbrites cause the valley slopes to be steep or nearly vertical, as is clearly shown by Photo 27 of the Ngarai Sianok valley (Water Buffalo Canyon) near Bukittinggi. Schmidt (1933) states that thin slices of the ignimbrites may collapse near escarpments but that building activities at short distances from the steep slopes are safe if the angle is 76° or less.

The intramontane plain is drained in two directions, so that a divide runs over the middle of the plain. The larger eastern part of the ignimbrite plain is drained by the Agam River, which turns eastward where it reaches the older folded mountains; after a short underground course in Permo-carboniferous limestones, it enters a *cluse* type of gorge mentioned by Van Valkenburg (1921). In the western part of the intramontane plain the ignimbrite extends at least 15 km NW of Bukittinggi in the Median Graben.

Photo 28. Vertical aerial view of a part of the Median Graben S of Sipirang (N of Bukittinggi). Scale 1 : 20,000. Note how the location of the small patches of alluvial plain (rice fields) and the drainage system are influenced by a fault running from left to right.

Photo 29. Vertical aerial view of the eastern fault scarp of the Median Graben near Lubuksikaping. Scale 1 : 20,000. Note how the straight scarplet at the foot of the large fault scarp influences the river course, which is deviated by landslides, such as the one at the extreme left.

8. *Masang-Sumpur Section*

The graben section NW of Bukittinggi is drained by the Masang River which, after leaving the graben, breaks through the western block mountains near Sipirang and debouches in the Indian Ocean. The tuff cover in the graben becomes gradually lower toward the NW.

Aerial photographs reveal interesting details of the faults occurring in the graben. Photo 28 shows a straight young scarplet S of Sipirang (34.2 x 2.7), where the graben is narrow; Photo 29, showing the eastern fault scarp near

Fig. 29. Field sketch of the tuff terraces in the Median Graben near Lubuksikaping, central Sumatra; looking NW. The Sorikmerapi volcano is visible in the distance.

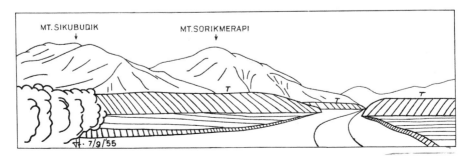

Lubuksikaping, demonstrates how the Sumpur River is influenced by the fault. Graben deposits forming terraces also occur here (Fig. 29).

The complicated fault pattern of this area, with fault directions deviating from the Sumatra direction, will be discussed in the chapter on northern Sumatra.

d. The Mountainous Areas to the East of the Median Graben

1. *The southern parts*

The mountains E of the Median Graben of southern Sumatra end at the northern end of the Gumai Mountains. The description of what follows to the N is more difficult, because the mountain complexes become much broader and are crossed by many rivers, so that they actually consist of numerous separate ridges and ranges. Accordingly, the larger complexes have no local names and have been designated by their explorers, mostly geologists, by the names of the rivers by which they are crossed or bound. We distinguish, for instance from S to N, the Tembesi-Rawas Mountains, the Tebo-Tabir Mountains, the Gumanti-Liki Mountains, and the Lisun-Kwantan Mountains. Although these names are rather cumbersome, they have the advantage that the areas can be found on any map containing the river names.

The first section, crossed by the Rawas River and becoming broader until it ends at the Tembesi River, is partly block-faulted and, like the Gumai Mountains, ends rather abruptly at the border of the eastern peneplain.

Some extinct and eroded volcanoes reaching heights of 2,156 and 2,383 m occur near the Median Graben. Their fluvio-volcanic fans have spread out over part of the older mountain area. A larger group of volcanoes situated further N contains both extinct cones with calderas and active ones with solfataras and crater lakes. This group is dominated by the beautiful cone of Mt Masurai (2,933 m), whose slopes carry some secondary craters with lakes. A number of recent cones and crater lakes, partially surrounded by a crescentic caldera wall, extend in a northwesterly direction.

The fluvio-volcanic products of these volcanoes spread freely toward the peneplain in the E through a gap between the Tembesi-Rawas Mountains and the next complex, the Tebo-Tabir Mountains. Various rivers flow through this gap. The lateral ones (e.g. the Tembesi River) have been pushed by the fluvio-volcanic flows against the bordering mountains, whereas the Merangin River, coming from the Kerintji graben, occupies a central position within the gap.

The next mountain complex, the Tebo-Tabir Massif, contains various granitic areas and rises much higher, to over 2,000 m. Near the Kerintji Section of the Median Graben it has the character of typical block mountains and includes some old extinct volcanoes.

Photo 30. View looking SE from the slopes of Mt Kerintji. The Mt Rubuh horst, separating the Median (Kerintji) Graben from the Bento marshes, can be seen in the centre. The Kerintji graben and Lake Kerintji can be seen at the far side of the mountain. The Bento marshes at the foot of Mt Rubuh are drained by the Sangir River seen at the left. The Kaju Aro tea estate is visible at the lower right. (Photo by Jacobs).

2. *Kerintji and Tudjuh volcanoes and related fluvio-volcanic features*

The Tebo-Tabir Mountains end at another wide gap through which the fluvio-volcanic flows of the Kerintji and Tudjuh volcanoes reach the eastern peneplain. These flows pushed the Tebo River against the mountain border.

The slender cone of Mt Kerintji (3,805 m), the highest peak of Sumatra, is certainly the most impressive of the volcanoes in the western end of this 'gap'. In the top area a deep crater, its bottom lying 3,442 m above sea-level, forms the centre of the volcanic activity. Two small parasitic cones and a crater lake are situated on the lower southwestern slopes, and there is another parasitic cone in the N. The tuff slopes are remarkably little dissected in the S, where the most recent deposition clearly occurred. The other slopes are distinctly more intensely dissected.

The southeastern slopes spread freely into the large Bento marshes (42.5 x 10.5; Photo 30), situated in a local depression or graben which is separated from the Median Graben by a narrow horst already mentioned. On these lower southeastern slopes, where the Kaju Aro tea estate is situated, the

Photo 31. The Kerintji lahar terraces as seen from the Kaju Aro estate. The Kerintji volcano and an adventive cone can be seen in the distance.

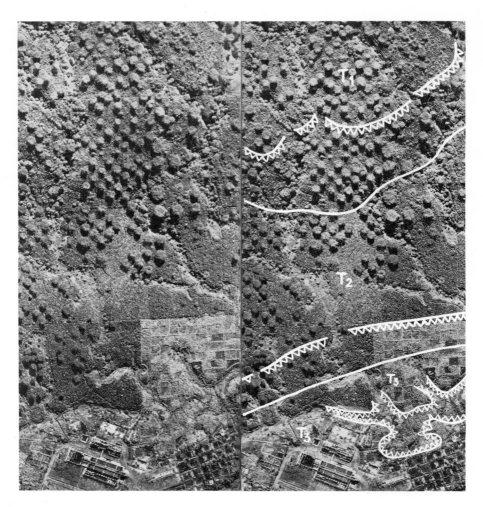

Photo 32. Stereogram of the Kerintji lahar terraces (scale 1 : 12,000). Three levels can be clearly distinguished, the buildings of the tea estate being located on the lowest one (T 3). Note the shade-trees among the plantations.

fluvio-volcanic flows can be seen to form three distinct levels. The buildings of this tea estate are situated on the lowest of these levels (1,450 m), as can be seen from the ground view in Photo 31 and the stereogram of Photo 32. The highest level lies at an altitude of approximately 1,550 m. The Kerintji volcanic products also spread northwestward in the Median Graben.

The extinct Mt Tudjuh (2,604 m) is a large caldera filled by a lake (Photo 33), whose surface is at 1,996 m above sea-level.

Photo 33. Looking E from Mt Kerintji, the volcanic Tudjuh Lake, lying 1,996 m above sea-level, can be seen at the left and Mt Uludjudjukan is in the distance behind the lake. (Photo by Jacobs.)

3. *Mountains between the Kerintji and Marapi volcanoes*

The southern half of this range is sometimes referred to as the Gumanti-Liki Mountains and is drained by the Batang Hari and its many tributaries. Valleys parallel to the Median Graben point to the existence of longitudinal faults. The heights gradually decrease toward the E, and a wide hilly area of complex structure, with heights around 600 m, forms the transition to the folded Tertiary basin.

Continuing the survey to the N, one arrives at the area E of Lake Singkarak

Fig. 30. Field sketch of the Kapitau gorge of the Silungkang River where it breaches the Palaeogene beds; looking W. The peneplain formed in the Neogene zone appears in the foreground, the diabases and granites of the Barisan Range in the distance.

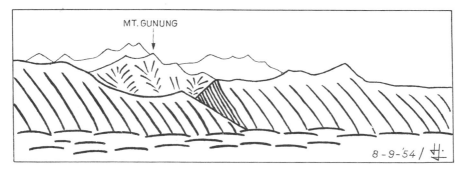

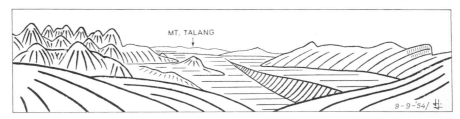

Fig. 31. Field sketch of the longitudinal depression S of Pajakumbuh, looking SE. Conical karst hills are seen on the left and Neogene beds fill the depression.

in the Median Graben. Some marked divergences from the eastern mountains hitherto described become apparent.

Firstly, it becomes difficult to maintain the distinction between complex mountains and block mountains made further to the S. Volcanism also becomes considerably less important.

Secondly, lower zones, extending in the Sumatra direction, occur here wherever rocks of poor resistance, mostly of Neogene age, are found. The Palaeogene-Neogene boundary often marks the sharp transition from the mountains to the longitudinal depressions. Fig. 30 illustrates this from the example of the Silungkang River entering the Neogene to the E of Sawahlunto. A broad intramontane depression marking the western limit of the folded ridges extends, for example, northward to Pajakumbuh (35.5 x 14; Fig. 31). E of this depression, the folded Palaeozoic rocks form prominent longitudinal ridges indicated as folded ridges on the geomorphological map. These mountains are referred to in the geological literature as the Lisun-Kwantan-Lalo Mountains. They are crossed by the Kwantan River; its tributaries, in accordance with the fold ridge relief, show a trellis-type drainage.

Well-developed conical karst areas, mostly in rather narrow belts and ridges, occur in the limestone zones, such as those depicted in Photos 34 and 35.

Photo 34. High conical karst hills in the eastern part of the Barisan range between Sungaidareh and Sidjundjung.

Photo 35. Conical karst hills SE of Batusangkar, looking E.

4. *Marapi and Malintang volcanoes*

Two young volcanic cones, i.e. the Malintang (2,262 m) in the E and the Marapi (2,891 m) in the W, crown the Barisan Mountains further to the NW. The Agam River loops around the northwestern, northern, and eastern feet of the Malintang volcano, whose volcanic products have pushed it against the mountains to the E. No eruptions are known for this volcano. The Marapi, on the contrary, is very active, and numerous eruptions of the explosive type are recorded. The craters are aligned along an ENE–WSW line within an old caldera, a remainder of which occurs in the northeastern part of the top area. The volcanic activity was gradually displaced towards the WSW.

It may safely be assumed that the Bukittinggi area once drained to the southeast toward the Anai gorge or Lake Singkarak but was later blocked by the further activity of these two volcanoes.

A WNW–ESE-running fault is an interesting feature of the northern foot of the Marapi volcano. This feature is so conspicuous that it is even clearly visible on the 1 : 40,000 topographic map (sheets 21 and 56). The country S of the fault has recently been slightly uplifted, forming a distinct young scarplet. The stereogram of Photo 36 depicts part of this scarplet to the E of

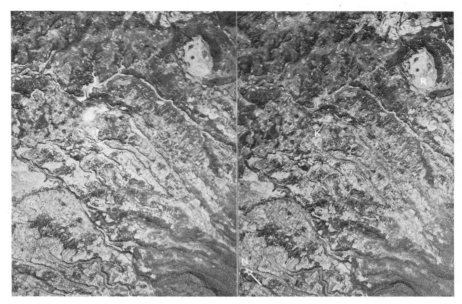

Photo 36. Stereogram of the lower northern parts of the Marapi slopes, central Sumatra. Scale approximately 1 : 75,000. Note the roughly WNW-ESE striking fault in the mantle of this volcano near Baso, indicated by the scarplet, the deviations of the various branches of the Agam River, and the location of two explosion craters, i.e. Telaga Kurai (K) and the larger one indicated by the rice-fields near Pulau (R), the latter being drained by the Bagurai River.

Fig. 32. Field sketch of a recent fault scarp (F) in the lower northern slopes of the Marapi volcano in central Sumatra E of Baso. Compare with Photo 36.

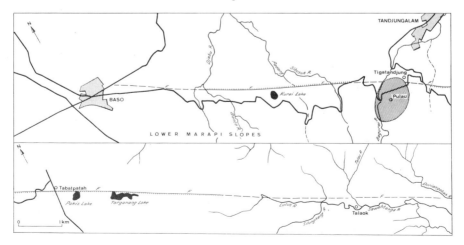

Baso. The rivers draining the slopes of the Marapi volcano (top left) show distinct offsets where they reach the fault scarp. They finally empty into the Agam River, beyond the area shown in the photo. The explosion craters formed along the fault, which evidently acted as a volcanic vent. The smaller of the two (K on the photo) is filled by a lake, the larger one (R) is occupied by rice fields (Fig. 32). Further to the E the fault runs in a slightly different direction, and the higher and more eroded scarp occurring there points to downthrow of the southern side.

5. *The northernmost parts*

Folded sedimentary rocks outcrop again to the NW of the Marapi volcano. As already mentioned (p 80), the Agam River, which drains the intramontane plain of Bukittinggi to the E, traverses this area, and after a short subterranean course enters a rather uncharacteristic *cluse* described by Van Valkenburg (1921). The mountainous zone now becomes very narrow, and is rather poorly known. A few old extinct volcanoes occur, but further to the NW there is no sign of volcanism.

e. The low areas of eastern Sumatra

1. *Peneplain*

This peneplain is rather similar to the one described in southern Sumatra, and therefore a brief description will suffice. The drainage pattern is largely of the trellis type. Because this is obvious even in the alluvial plain, it is evident that recent tectonics contribute to this characteristic drainage pattern. A Sub-Barisan depression often marks the boundary with the Barisan Mountains.

Within the low areas there are a few higher areas where more intense folding occurred; these include, for example, the Duabelas Mountains and the larger Tigapuluh Mountains, whose highest parts are formed by a number of ridges composed of more resistant pre-Tertiary sandstones and graywackes. An area of considerable relief located within the belt of folded mountains E of Lubuksikaping and drained by the Kampar-kanan River is shown in Photo 37.

Old volcanic tuffs cover parts of the peneplain, particularly near the outlets of the gaps in the eastern Barisan Mountains. These are not the only occurrences however, and there are also tuffs in areas with piedmont remnants showing little or no erosion. These areas, too, are often characterized by a radial drainage pattern.

River terraces are well developed in the valleys draining the peneplain. Terraces at heights of 6, 11, and 23 m above the Merangin River were observed, for example, near Bangko. Terraces with relative heights of 3, 7, 13 and 17 m occur along the Batang Hari where it enters the lowlands near

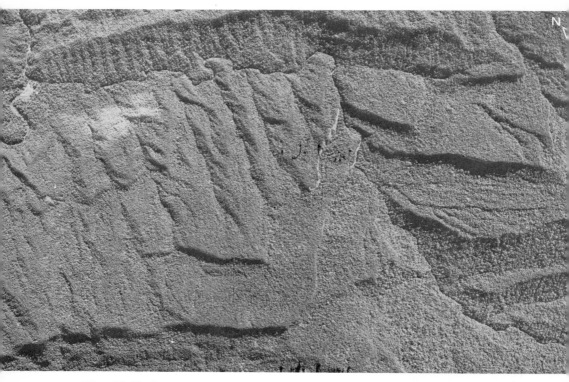

Photo 37. Vertical aerial view of part of the folded mountains of eastern Sumatra to the east of Lubuksikaping, in the Kampar Kanan River basin (scale about 1 : 20,000). The distinct dip-slopes of the more resistant beds and alternating subsequent and consequent river sections are characteristic.

Sungaidareh. The 3 m terrace and some relics of the 17 m terrace were also observed further upstream, where the Batang Hari flows through an eroded mountainous terrain.

2. *Drainage pattern South of the Kwantan River*

A swampy alluvial area located where the Merangin River empties into the Tembesi River deserves special mention. Obviously, this area is structurally controlled and is comparable to the swamp drained by the Rupit and Rawas rivers further to the S described on p 56. The depression extends in the W as far as the Tabir River and runs parallel to the fold areas of the Duabelas Mountains (431 m).

The large Batang Hari also flows in a longitudinal depression to the SE, parallel to the Tigapuluh Mountains. The NE-running downstream part breaks through the anticlinorium where the latter is strongly depressed. The Kwantan River occupies a similar position NW of these mountains.

In general, the distribution of alluvial plains and swamps and their varying width throughout eastern Sumatra yields important information about the location of zones of upheaval and subsidence. The Batang Hari delta is indicated as alluvial lowland on the geomorphological map, in accordance with the topographical maps. The geological map 1 : 1,000,000 indicates a rather extensive occurrence of old volcanic tuffs. The new 1 : 2,000,000 geological map of Indonesia (1965), however, gives a divergent picture, indicating that volcanic products have been transported far into the lowland zone by rivers flowing through the gaps in the Barisan Mountains.

3. *Drainage pattern North of the Kwantan River*

The longitudinal drainage is considerably better developed N of the Kwantan River. This is particularly distinct near the Barisan Range, where higher folded mountains occur. The upper reaches of the Kwantan, Kampar-kiri, Kampar-kanan, and Rokan-kiri rivers therefore show a typical trellis drainage pattern.

The Kwantan River further downstream has two sections flowing approximately parallel to the fold axes. The first continues to the NW in the broad swamps of the Teso River belonging to the Kampar-kiri drainage basin, and can be traced even further in this direction.

The swamps of the Nilo River, also draining to the Kampar-kiri, offer another example further downstream. The longitudinal Kampar-kanan section, immediately upstream from its junction with the Kampar-kiri, is situated within the same zone. It continues in the swampy dry valley located SE of Pakanbaru and forming a connection between the Kampar-kanan and Siak valleys. The upper Siak River may once have formed part of the Kampar-kanan drainage basin. The lower Tapung-kanan River marks the northwestern continuation of this zone. Neotectonics probably played a role in the important changes in the drainage of this area.

Further to the NW we arrive at the Rokan-kiri drainage basin, where the alluvial plain penetrates far inland and where the zoning of the peneplain can no longer be traced.

4. *Recent tectonics in the lowlands*

Returning to the Kampar River, we find that part of its lower course is distinctly longitudinal, as is its left tributary the Palalawan River. The Mandau branch of the Siak River and the Kampar River near Rengat are situated in the same zone. Because the longitudinal sections of the Siak and the Kampar rivers are located in the alluvial plain, only young tectonics, i.e. recent subsidence, can account for them. It is likely that Neogene rocks occur at shallow depth (see geological map 1 : 2,000,000, 1965).

Recent subsidence also seems indicated for a longitudinal zone of rivers with swamps or lakes in the alluvial plain further to the E. The Serkap River and the associated lakes and swamps mark the southeastern part of this zone. The situation is even more distinct further to the NW, where the Siak

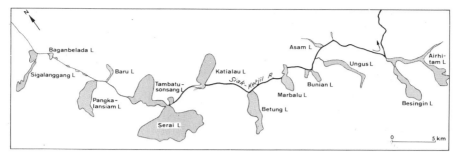

Fig. 33. Lakes formed along the Siak Ketjil River due to recent subsidence. After the 1 : 100,000 topographical map S.O.K., sheets 21/XV + XVI and 22/XV + XVI.

Ketjil River is bordered by many lakes (Fig. 33). The lower course of the Rokan River enters into the same zone.

On this basis it seems probable that the peninsula of Bagansiapiapi coincides with a zone of young tectonic uplift, and the same may apply to the islands (Tebingtinggi, Bengkalis, Rupat, etc.) bordering this part of the E coast. The Pandjang, Airhitam, and Rupat straits may very well represent zones of recent subsidence comparable to the synclinoria occurring further to the W. The zone where maximum subsidence occurred during the Neogene and in more recent times, is located slightly inland, however, since the coastal areas are too close to the Sunda Shelf. The drainage pattern of the eastern lowlands of central Sumatra certainly invites a geomorphological investigation.

5. *Beach ridges*

A number of well-developed beach ridges occur on some of the above-mentioned islands. Their materials probably derived from the submerged parts of the old Sunda land, of which the nearby Kundur and Karimon islands represent emergences. Beach ridges of any importance are otherwise lacking in the broad alluvial plains of Central Sumatra. This is easily explained, because the sandy portion of the river-load is deposited upstream and never reaches the sea, a very natural situation in a humid tropical environment where intense weathering produces predominantly clayey waste material.

VII Northern Sumatra

a. Introduction

Geomorphologically, this is undoubtedly the least known part of the island. Topographical maps for Atjeh, i.e. the northernmost tip of the island, are lacking or inadequate. Little geological work has been carried out, except in the oil fields in the northern part of the eastern geosyncline. The geology of the Lake Toba area has been studied in some detail, and the present author has also carried out some geomorphological work in this interesting region.

There is a certain resemblance to the geomorphology of central Sumatra, at least as far as the block mountains to the W of the Median Graben and the graben itself are concerned. Intense block faulting occurred, but no distinct and consistent tilt of the blocks, which is so characteristic for the southernmost part of the island, can be observed. The differences in height between the blocks are even more pronounced locally than in central Sumatra. The highest block, Mt Leuser (3,381 m), occurs to the W of the longitudinal graben, whereas the alluvial plains of Meulaboh and Singkel indicate the position of subsided blocks along the W coast. The longitudinal graben can be traced through the whole length of northern Sumatra and even in the sea floor beyond the northern end of the island.

More complicated geomorphologically are the parts E of the Median Graben. In the S, near central Sumatra, these block mountains are narrower than anywhere else on the island. The Toba graben is situated further N, where a series of cataclysmic volcanic eruptions of the Katmaian type occurred, as a result of which an extensive ignimbrite plateau was formed, covering the underlying structures. The only K-Ar dating of the ignimbrites existing to date gives an age of $< 300{,}000$ years (Katili, 1969). Deep narrow steep-sided gorges are a major geomorphological feature of this plateau. Other young volcanic activity is also known for this area, in contrast to the adjacent sections to the N and S. A zone of well-developed, though rather narrow, block mountains occurs further to the N.

A new geomorphological element can be distinguished in Atjeh: broad basins change the character of the Barisan Mountains significantly. Another anomaly here is the deviation from the Sumatra direction of some major secondary grabens, such as the one of Lake Tawar, as well of as many other structural elements. However, the situation becomes quite normal again

near the northernmost tip of the island. Recent volcanism is confined to:
a) the row of volcanoes N of the Toba area,
b) the stratovolcanoes to the NW of Lake Tawar, and
c) the cones SE of Kotaradja.

The alluvial plain along the E coast becomes gradually narrower to the N. Sedimentation was evidently less intense here, or at least could not keep pace with subsidence. Thus, the Malacca Straits become broader toward the N and the island of Sumatra correspondingly narrower.

b. The Block Mountains to the West of the Median Graben

1. *Malintang volcanic complex (31 x 12.5)*

The southernmost parts of the western zone are widely covered by young volcanic products. The volcanic mudflows of Mt Malintang (1,983 m) spread far to the W and SW, as already pointed out earlier, and border the lowlands of Airbangis on the landward side. A large crater lake, measuring about 900 x 1500 m, is found in the top part of this volcano. The extensive fluvio-volcanic flows just mentioned are best accounted for by assuming that this crater lake was emptied on several occasions during prehistoric eruptions. NW of the volcano there is a curved and well-dissected ridge probably representing a relic of an old caldera, open to the S. Further to the N and NW there are some smaller and severely eroded cones of extinct volcanoes.

2. *Sorikmerapi volcano and adjoining faults*

The northernmost of this group of volcanoes is the active Mt Sorikmerapi (2,145 m). This volcano has been described at length by Kemmerling (1920) and will therefore only be dealt with briefly here. Photo 38 shows part of the

Photo 38. Part of the Sorikmerapi caldera bottom, looking to the S. Note the mud-well on the left and the sulphuric mud deposits in the foreground.

Fig. 34. Field sketch of the Sorikmerapi volcano with the fault scarp (F) on its eastern flank, as seen from the graben at Sibonggartengah, looking SW.

large deep caldera bottom, where volcanic mud wells have produced sizeable sulphur deposits. Two small mud wells indicated on Kemmerling's crater map had disappeared by the time the author visited the volcano in September 1955. Solfatara occur on the southern slope of the caldera interior and on the lower part of the outer eastern slopes, to the left of the trail leading to the top. A number of trees died off here when poisonous gases escaped in 1951.

The real danger of the Sorikmerapi lies not so much in its volcanic activity but rather in the occurrence of a fault together with some minor grabens, running right across its eastern slopes. Major landslides, such as the one of 1915, are provoked by the continuing activity of this Sibonggar fault zone. Figs. 34 and 37 show the situation.

The uplifted eastern limb of the fault is formed by the Maga Plateau, sloping gently to the E and extending into the Median Graben. This feature should be regarded as an old lahar foot of the Sorikmerapi that developed before the above-mentioned fault zone came into being (Fig. 35, Photo 39). Its coarse volcanic agglomerates were deposited by mudflows (Kemmerling, 1920). The tiny Sibonggar graben, where sulphur mud pools occur, forms part of this fault zone whose presence at the foot of the Sorikmerapi has a

Fig. 35. Field sketch of the Padangsidempuan section of the Median Graben (1) looking NW from a point W of Hutanopan. The volcanic Maga Plateau (2), originating from the Sorikmerapi volcano, covers the graben in the foreground and reaches the eastern fault scarp (cf. Photo 39).

Photo 39. The graben of Padangsidempuan looking NW from a point approximately 7 km W of Hutanopan. The eastern fault scarp and the volcanic Maga plateau originating from Mt Sorikmerapi are clearly visible.

Fig. 36. Sketch map of the area between the Sorikmerapi and Hutanopan mountains (for location, see Fig. 38).

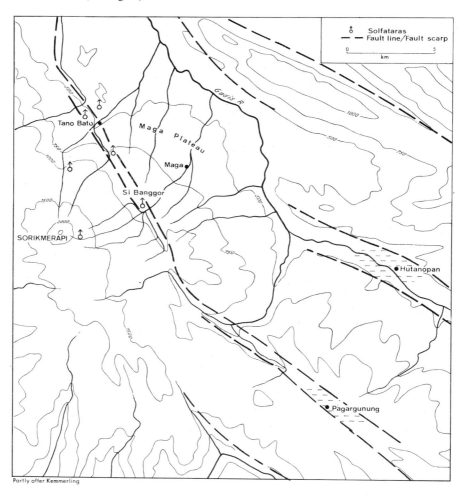

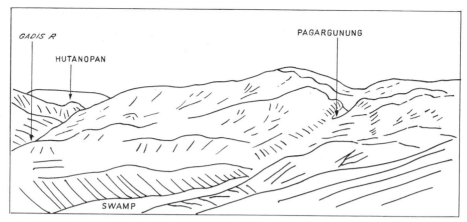

Fig. 37. Field sketch of the high-lying graben section, with the Pagargunung plain, located SE of Mt Sorikmerapi, as seen from the top of the volcano. The graben becomes lower and broader in the left foreground. The graben of Hutanopan with the Gadis River can be seen on the left.

pronounced influence on the volcanic morphology, drainage, etc. All but three of the small streams draining the eastern Sorikmerapi slopes are diverted by the fault scarp (Kemmerling, 1920). A number of 'wind gaps' indicate the sites of their original downstream parts (Fig. 36; for location, see Fig. 38). The high position of these torsos provides evidence of the activity of the fault in more recent times.

The continuation of the fault can be traced further to the SE, where it influences the courses of several small rivers. Here, it distinctly takes on the character of a graben. The village of Pagargunung is situated on a high isolated and elongated small plain (Fig. 38, Photo 40). A terrace lying about 10 m above the present sawah plain was formed by the incision of the Siabut

Photo 40. The small, high graben of Pagargunung seen longitudinally, looking SE. Note the 10 m tuff terrace in the distance and the low saddle separating this plain from the continuation of the graben.

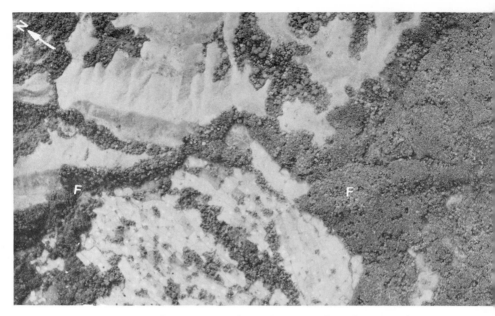

Photo 41. Vertical aerial view of the Tanobato fault, situated N of the Sorikmerapi volcano. The young scarplet (F–F) is clearly visible and its influence on the drainage evident. Scale about 1 : 15,000.

River. The deposits cropping out in the terrace scarp show alternating layers of fine and medium tectured volcanic tuffs. Particularly interesting is the occurrence of three thin intercalated peat layers, which suggests that the repeated filling of the graben by volcanic (Sorikmerapi) products occasionally alternated with further subsidence of this part of the graben, hampering its drainage and giving rise to a swamp or lake subsequently filled up again by volcanics. Since the Pagargunung plain is drained in the direction of the graben, the phenomena observed must be due to differential subsidence within the graben and cannot be ascribed to the relative uplift of the neighbouring horsts.

The fault scarp can also be traced NW of the Sorikmerapi, toward Tanobato (Photo 41). A young scarplet having a clear impact on the drainage can be recognized. Several rivers break through the higher eastern limb. The Sorikmerapi lahar flows have clearly been guided here by the fault topography, thus providing evidence that the fault existed before the most recent main volcanic activity started. This is not surprising, since the fault may be regarded as the southern continuation of a fault bordering the Padangsidempuan section of the Median Graben. The various grabens and faults occurring in this part of the Barisan Mountains makes it difficult to determine which should be considered to represent the main Median Graben. It

is usually assumed that it has two branches here, the broad Hutanopan and Sumpur grabens. The present author agrees with this view, although as far as its position is concerned the Pagargunung-Sibonggar graben might equally well be considered a branch of the Median Graben.

3. *The Natal plain and its hinterland*
Extensive swampy lowlands occur W of the volcanic complex just described. The town of Natal is located on the coast, along which a few rocky, mostly andesitic capes occur. A submerged block probably forms the subsoil of the swamp, in much the same way as suggested (p 65) for the adjoining Airbangis plain. It should be noted here that a largely submarine swell, crowned by Pini Island, connects this part of Sumatra with the non-volcanic arc. No signs of abrasion of the coast that would point to subsidence in recent times were observed, however.

On the landward side of the swampy plain, and also in the rocky areas closer to the coast, there is a rather well-bevelled surface lying about 75 m above sea-level. This surface is well developed in the rocks having little resistance, but harder ridges still stand out in relief and thus influence the drainage, e.g. of the Natal River. Terraces lying 2 and 6 m above the river and also relics situated about 25 m above the river can be seen in this valley, but higher terraces are absent. A good part of this bevelled terrain is covered by volcanic tuffs on which a characteristically dentric drainage pattern has been formed.

4. *Strongly block-faulted areas between Natal and Tapanuli Bay*

A rather narrow alluvial plain continues along the coast until it widens near the southern shore of Tapanuli Bay (Photo 42). A number of isolated hills and ridges stretching in the Sumatra direction emerge from this plain; some of these are andesitic. The island of Musala also belongs in this class.

The drainage of the hinterland is largely of the angulate type, and many rivers flow in the Sumatra direction at the foot of steep straight ridges. Faulting is obvious here in several places. The small Si Ais Lake (25.5 x 12.5), for instance, is formed in a minor graben which disappears under the alluvium toward the NW. Much larger is the swampy graben found further to the SE. This graben is drained by the Gadis River, which crosses the western fault scarp of the Median Graben and the block mountains. It is interesting to note that the southernmost extremity of the swampy graben is drained by a parallel river, so that a very shallow divide is formed in the swamp. This graben might be turned into an agricultural area if the drainage could be improved.

This part of the western block mountains is cut off in the N by a branch of the Median Graben running from Padangsidempuan in a WNW direction and disappearing under the alluvium S of Tapanuli Bay.

5. *Hinterland of Tapanuli Bay and areas West of Lake Toba*

The next section of the western mountains, the Sibolga Block (approximately 600–1,000 m) occurs inland from Tapanuli Bay. The lahar products and ashes of the Lubukraja volcano reach far westward between this block and the previously-mentioned western branch of the Median Graben. These volcanic products effectively push the Batang Toru against the Sibolga Block (26 x 14/15). The latter is mainly composed of granites with some occasional quartzites and shales. An anomaly within this block is the radial drainage

Photo 42. Tapanuli Bay with the town of Sibolga, as seen from the road to Tarutung.

of Mt Dolokgindjang, which points to a different lithological composition and a volcanic origin of this mountain.

To the N of Tapanuli Bay a rather narrow alluvial plain is again found separating the mountains from the sea. The mountain face is straight and runs in the Sumatra direction. All settlements are located at the foot of the mountains on the landward side of the plain.

A narrow strip near the eastern edge of the block mountains in this part of Sumatra is covered by the Toba ignimbrite plateau, from which only a few isolated mountain peaks emerge. The ignimbrites are mostly limited by a straight low ridge running in the Sumatra direction. To the NW of the lake the ignimbrites penetrate into the Sembean valley, which is probably a tectonic valley running at a slight angle to the Median Graben.

6. Singkel – Meulaboh Section

The alluvial plain, narrow at first, suddenly becomes considerably wider as it approaches Singkel. This part of the plain is actually part of a largely submarine ridge, crowned by the Banjak Islands and connecting Sumatra with the non-volcanic arc further to the W. The plain is traversed by the Simpang-kiri and Simpang-kanan rivers, both draining a large part of the western block mountains, which have not received much study in this section. Low hills of Tertiary rocks form a transition from the plain to the mountains. The narrow NW extremity of this plain, where numerous beach ridges occur, finally ends where the mountains reach the sea again. In the hinterland the highest part of the western block mountains occurs in the Mt Leuser (3,381 m) area. Only one volcano in Sumatra, Mt Kerintji, reaches higher.

The rocky coast is probably a fault coast, and further NW confirmation of the fault is found where a zone of alluvial lowland is located at its foot. A vertical aerial view of this straight steep mountain face is given in Photo 43. It can be seen that one river breaks through the fault, whereas another

Photo 43. Vertical aerial view of an important fault scarp striking in the Sumatra direction and forming the edge of the Meulaboh alluvial plain near Babahrat. The small Sapi River breaks through the scarp and enters the plain; the Batee River runs at the foot of the escarpment. Scale about 1 : 15,000.

runs along in front of it. This alluvial plain, in the middle of which the town of Meulaboh is located, probably owes its existence to a submerged block underlying it at shallow depth. A series of beach ridges occurs all along this part of the coast.

7. *Northernmost parts of the western block mountains*
Tertiary hills form the hinterland of the plain over a considerable distance, and the block mountains are narrow in these parts. Van Bemmelen (1949) assumes the existence of transverse faults in this sector, one running along the Tripa River and the other along the Teunom River.

In the final section the rocks reach the coast again. Lofty mountains reaching almost 2,000 m occur here. In some limestone areas, caves and beautifully developed conical karst hills have been formed, for instance SW of Kotaradja in a fault valley with a blind end.

The western zone of block mountains terminates in two rocky islands and a former island now united with the mainland of Sumatra by an alluvial plain. Their rectilinear eastern flanks apparently represent the fault scarp bordering the Median Graben.

c. The Median Graben

1. *The transition zone toward central Sumatra*
In the area N of the Padang Highlands the fault pattern becomes so complicated that the location of the Median Graben (Fig. 38) becomes a problem.

First, there is the broad graben N of Lubuksikaping, in which the Sumpur River flows before it breaks through the eastern mountains (Photo 44). This graben, however, takes an almost northerly course, thus deviating more and more from the Sumatra direction and ending in the eastern chain of the Barisan Mountains. Since it does not occupy a median position in this narrowest section of the whole mountain range, its continuation will be discussed in the paragraph dealing with the mountains to the E of the Median Graben.

Secondly, there is the high pass leading from the northern end of the Sumpur Graben in a northwestern direction to Hutanopan at the southern end of the next graben section, i.e. that of the Gadis River. Although it has a straight course and clearly follows a fault along the upper Gadis River, it hardly has the character of a true graben. On the other hand, the graben of Pagargunung in the western chain, treated in detail on p 97, is small and its southern end lies too far outside the heart of the mountain range to be considered Median Graben.

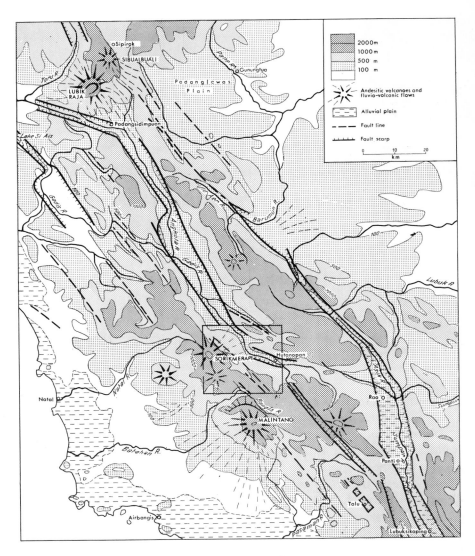

Fig. 38. Sketch map of the southwestern part of northern Sumatra.

A more logical solution would therefore to be trace the Median Graben in this part of the island from a fault W of the Lubuksikaping graben to a narrow but distinct graben (31/32 x 13) in the western mountains, drained further on by the Pungkut River to the NW. The latter river makes a sharp turn to the NE and enters the Batang Gadis, which runs through the plain of Hutanopan. This is again a graben structure, although somewhat broken

Photo 44. Downstream view of the valley of the Sumpur after the river has left the Panti-Rau Section of the graben. Here, the river breaks through the fault scarp and the block mountains to the E of the graben. The absence of a flat valley bottom and terraces points to the predominance of vertical erosion in this part of the river.

up due at least partially to the occurrence of granitic intrusions*. The Median Graben, finally, is clearly established again in the broad graben of Padangsidempuan.

It must be concluded that the localization of the Median Graben in these strongly block-faulted areas is open to discussion. It would perhaps be more appropriate to say that the Barisan geanticline here is intensely faulted without the formation of one clearly defined Median Graben. Two main graben zones exist, i.e. the one just described and the Sumpur graben (Photo 45) close to the eastern flank of the Barisan (Fig. 38).

* A comparable situation has been mentioned for Kerintji (p 70), where a granite massif affected the development of the Median Graben, and the area N of Padangsidempuan, where the Sibolga granitic intrusion seems to have played a role in the genesis of a bifurcation of the Median Graben.

Photo 45. The broad Sumpur graben, looking NW from a point 2 km E of Rau. Graben deposits form terraces on either side of the graben.

2. *Deposits and terraces in the graben*

Near Hutanopan, terraces occur along the Batang Gadis at heights of approximately 6 and 25 m. Their relative heights remain unchanged further downstream, and the inclined surfaces thus produced are proof that they cannot be explained as 'lake deposits', as suggested by several earlier authors attempting to account for such graben deposits. The nature of these deposits varies greatly from one locality to the other. Volcanic tuffs and coarse lahar deposits occur, but pebbly and silty deposits have also been found. Even occasional peat layers occur locally. In the present author's opinion, these deposits, which are always flat-topped, point to a period of subsidence with associated deposition followed by river incision and erosion. The lower parts of the scarps bordering the graben should in these cases be considered as fault-line scarps, whereas their higher portions are true fault scarps. We shall return to this problem further on (p 107).

The 6 and 25 m Batang Gadis terraces can also be traced where this river is incised in the coarse lahar deposits of the Maga Plateau (Fig. 35), which originate from the Sorikmerapi volcano (p 95). The Hutanopan plain was thus blocked by these volcanic products and the Batang Gadis was pushed against the eastern fault scarp. The river then runs in a broad part of the graben lying in an alluvial plain, until it is pushed against the western fault scarp by a rather large fan. The latter may well be of volcanic origin, although its source is obscure.

3. *Padangsidempuan Section*

The Gadis River breaks through at a rather low place in the western fault scarp. The Angkola River, coming from the opposite direction, joins the Batang Gadis before the latter leaves the plain (Fig. 38). Photos 39 and 46 show this part of the Median Graben.

An interesting area is found along the Angkola River about 19 km upstream from this junction; here, the Median Graben seems to become nar-

Photo 46. The (cloud-covered) Padangsidempuan section of the Median Graben, seen looking NW from the top of the Sorikmerapi volcano.

Photo 47. Tuff terraces in the Median Graben SE of Padangsidempuan. Some facetted spurs are faintly visible along the eastern fault scarp in the distance.

rower (Photo 47). The eastern fault scarp is well developed here and rather undisturbed. Facetted spurs can be seen at its foot. The western fault scarp is rather obscure, however, and a small local horst seems to occur here (Fig. 39). This flat-topped feature is situated at a height of 450–470 m above sea-level where Tertiary rocks of little resistance are exposed. The surface of these rocks is covered with a whitish sediment, rich in quartz. These are probably ordinary graben deposits, although in one place deposits conceivably representing a delta structure were observed. Thus, the possibility that a lake once existed in the Padangsidempuan graben section cannot be excluded. A low (2 m) terrace is found where the Angkola River breaks through this local horst separating the alluvial plain of Padangsidempuan from that further to the SE.

The eastern fault scarp bordering the graben is clearly visible – both in the field and on the aerial photos – E of Padangsidempuan, where it forms a continuous escarpment. Some low hills represent less deeply subsided graben portions. Here, it is again evident that both river work and tectonics contributed to the development of the graben morphology.

Fig. 39. Field sketch of part of the eastern fault scarp of the Padangsidempuan section of the Median Graben with facetted spurs and a local horst composed of graben deposits at its foot. The Angkola River can be seen in the foreground.

4. *The graben South of Lake Toba*

The Angkola plain terminates where the extinct Lubukraja stratovolcano rises NW of Padangsidempuan, thus obliterating the fault phenomena. The Median Graben can only be traced again at the other side of this volcano in the long and perfectly rectilinear valley of the Toru River, which has its source on the Toba ignimbrite plateau (Fig. 47). The Toba ignimbrites have penetrated far southeastward in this valley, and, due to subsequent activity of the graben, these deposits now form a distinct 'terrace' which can be traced all along this graben section (Fig. 40). This graben fill gradually becomes thinner downstream. Many localities show a cover of surface materials often reddish, washed down from the adjacent mountains.

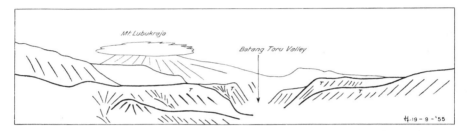

Fig. 40. Field sketch of the Median Graben about 10 km S of Tarutung, drained by the Batang Toru; looking SE. The Toba ignimbrites form a broad terrace (T) in the valley.

The Toru valley widens where the Puli River enters it from the E (25 x 15.5). A plain is formed here, the northern part of which is occupied by rice fields, the southern portion being swampy. The braided Toru River borders this plain on the W and the meandering Puli River on the E. This is a strongly subsided graben section where subsequent river erosion has not only largely removed the Toba ignimbrites but has also exposed younger graben deposits.

An interesting section of these deposits was studied on the eastern side of the valley. The outcrop is about 2.5 m high and 10 m long, and it lies roughly 35 m above the present river bed. The young unconsolidated beds are tilted at an angle of 38°, which makes it possible to study a rather thick sequence. This tilt is a completely local phenomenon; a little further on, the beds are horizontal. From top to bottom, the following layers were observed:

200 cm medium-textured light-coloured sands with numerous biotite flakes;
200 cm yellowish-grey clays;
  8 cm black peat, gradually passing into
 15 cm dark-grey clay with plant remains, passing into
  6 cm black peat;
 40 cm dark-grey clay;

15 cm black peat;
200 cm dark-grey clay;
40 cm dark-grey clay with peat;
80 cm light-grey clay with some plant remains;
15 cm blackish peat, passing into
40 cm greyish clayey peat, passing into
20 cm greyish-black peat;
35 cm finely-bedded sands and greyish-green clays with some plant remains;
30 peat
? cm medium-textured greyish sands, partly covered by a slump.

It is evident from this sequence that subsidence of the graben repeatedly caused swampy conditions, and alternated with filling of the graben, at least partially, with volcanic material. A similar situation has been mentioned for the small Pagargunung graben near the Sorikmerapi volcano (p 98).

Further to the NW the graben lies at a much higher altitude because the Toba ignimbrite fill is thick here and only partly eroded. In the alluvial plain of Tarutung (Photo 48) a broad terrace occurs about 50 m above the plain. Zwierzycki (1919) speaks of 'lake deposits' formed in a lake ponded-up by an andesitic outflow lower down in the graben. The Toru River carved a narrow gorge here, and the terrace terminates. The present author is of the opinion that these andesites are much older than the deposits considered here. He sees this terrace material as graben deposits formed upstream, NW of the pre-existing andesites, and now forming a terrace due to more recent river incision. It should be added that recent tectonics – more precisely, further subsidence of the Tarutung plain – also seem to have played a role. The straightness of the terrace edge and also the occurrence of hot springs, point to the presence of a fault which can also be traced further up-valley. The terrace decreases in height to 5 m and less about 20 km upstream.

Photo 48. The Tarutung section of the Median Graben with the Toru River as seen from the SE.

5. *The Median Graben West of Lake Toba*

The northwestward continuation of the Median Graben has been largely filled up by Toba ignimbrites. The same material forms an extensive plateau at a greater height on either side of the graben, from which, however, a number of mostly andesitic peaks emerge. Post-ignimbrite fault movements account for the straight courses of various rivers.

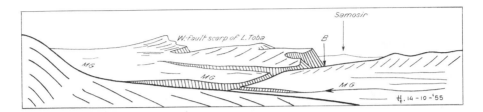

Fig. 41. Field sketch of the poorly developed and largely ignimbrite-covered Median Graben (MG) W of Lake Toba, looking NW. This graben section is drained toward the lake by the gorge of Bakara (B).

The Median Graben – still largely filled with Toba ignimbrites – then runs parallel to the W coast of Lake Toba, at a distance of only 12 to 15 km (Fig. 47). It is first drained by a small river which descends to Lake Toba through a steep gorge (Figs. 41 and 48). A gently sloping low hill (21.5 x 16) of unconsolidated ignimbrites forms a graben watershed, to the NW of which the Renun River originates. This fault valley is remarkably asymmetric, its western side being higher and considerably steeper than the eastern slope. Everything is covered by the whitish, unconsolidated young Toba ignimbrites, but lower down the older and more resistant Toba ignimbrites crop out in the valley, causing waterfalls in the river. The influence of a few minor faults on the drainage can be observed in air photos of this area. It may well be that this

Photo 49. The low broad Median Graben NW of Lake Toba, drained by the Renun River, looking SE.

Photo 50. Vertical aerial view of a young fault scarp on the western side of the Renun (median) graben S of the Lisang-Gunung capture. Scale about 1 : 30,000.

is not a true graben, but rather that only a single fault scarp is developed here along the upper Renun River.

The Renun valley rapidly becomes lower further to the NW, and finally a broad plain is formed at about 200 m (Photo 49). In one place the river escapes westward from this zone (18.5 x 16.5), to return to it a few kilometres further on. Over this stretch its small Belulus tributary occupies the depression. This feature is due to some peculiar fault phenomena. It appears from the air photos that only a single fault scarp is present here, as already mentioned for the upstream part of the Renun valley. The anomaly in the drainage is due to the fact that the downthrown side of the fault scarp, before the Renun leaves the graben, lies W of the river, whereas the E side is downthrown along the Belulus River. Photo 50 shows this clearly. The fault can be traced over a long distance, at least until the Renun River finally leaves the Median Graben zone to enter the western block mountains. No major scarp is formed there by this fault, but its impact on the drainage is evident.

6. *Kutatjane (Alas) plain*

The Toba ignimbrites can be traced in the Median Graben zone as far north-westward as the junction with the Gunung River coming from the E (18 x 16.5). A secondary fault can be observed here, deviating slightly from the Sumatra direction. A true graben thus appears again. The Gunung River basin drains part of the Toba ignimbrite plateau, and it displays an interesting capture, to be dealt with below (p 147). It should be noted here, however,

Photo 51. The Alas graben, looking NW from the same point as Photo 52. Note the terrace fragment in the middle of the graben in the distance.

Photo 52. Part of the western fault scarp of the Alas section of the Median Graben S of Kutatjane. The drainage formerly occurred to the left; terraced graben deposits can also be seen.

that the situation points to a considerable subsidence of this part of the Median Graben after the Toba eruption.

A broad plain at an altitude of only about 200 m, is found NW of the Gunung River (Photo 51) where the town of Kutatjane is located; this plain is drained by the Alas River coming from the NW after originating in the lofty Leuser area in the western block mountains. A true graben is developed here, and both faults bordering it can be clearly traced on the air photos as well as in the field. Photo 52 gives a ground view of part of the western fault scarp in this area. The vertical aerial view of Photo 53 shows a young scarplet formed at the foot of the eastern fault scarp S of Kutatjane. The height of the scarplet and the number of rivers crossing it being insufficient to allow the formation of triangular-shaped facetted spurs, trapeziform facets are developed instead. The offset of streams may point to a horizontal movement (arrows on the photo) along this fault. This is observed hardly anywhere else in Sumatra, a fact which – although certainly resulting in part from difficulties of observation – seems to point to a predominance of vertical movements within the Sumatran fault system.* The high well-dissected main eastern fault scarp separating the graben from the Serbolangit block rises behind this scarplet.

---

* It should be remembered in this connexion, however, that geodetic evidence for horizontal movements was gathered on one occasion, i.e. after the 1892 earthquake in Northern Sumatra. At that time, displacement of triangulation points was observed (Muller, 1895).

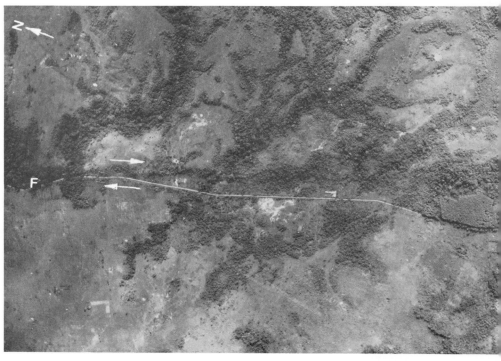

Photo 53. Vertical aerial view of a young scarplet on the eastern side of the Alas Section of the Median Graben S of Kutatjane. Scale about 1 : 17,500. Note the offset of the streams, possibly indicating horizontal movement along the fault indicated by arrows.

Fig. 42. Field sketch of the western fault scarp of the Alas Section of the Median Graben S of Kutatjane where the Alas River, which drains the graben, enters the western block mountains to debouch in the Indian Ocean.

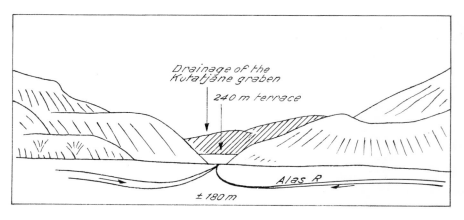

Deposits of sands and fine gravel are found in a terrace at about 50 m above the plain, i.e. 240 m above sea-level. Relics of this terrace occur on the eastern side of the graben just opposite the gorge where the Alas River leaves the graben to join the Simpang-kiri River, which debouches into the Indian Ocean. A field sketch of the situation is given in Fig. 42.

The drainage of this area poses an interesting geomorphological problem. The gaps through which the Alas and Renun rivers, coming from opposite directions, leave the graben, are 25 km apart, and the intermediate graben section is drained by neither of these rivers. In this section between the two gaps, however, a dry gap (wind gap) occurs where a fragment of the Alas terrace is found at 220 m above sea-level, now forming the divide with the Renun River. The reason why the Alas River abandoned this southerly outlet is obscure, and the question of whether a true lake once existed in this graben section cannot be answered with certainty either. Did the above-mentioned relic seen at 240 m near Laubalen once form a graben divide? This 240-m level ascends to about 310 m further to the NW where the Alas plain ends.

7. *Blangkedjeren basin*

NW of the Alas plain the graben becomes narrow and forested. Only the straight drainage lines indicate the position of the fault(s). Solfatara occur along the road to Blangkedjeren, about 47 km NW of Kutatjane. Approximately 20 km further on, the tiny basin of Gumpang (14 x 17.5) is located, but the graben only widens again definitely where the Blangkedjeren basin begins, about 8 km SE of this isolated town. This basin is one of a number of basin structures characterizing the northernmost part of the Barisan Mountains (p 93). It is drained by the Tripa River which, after making a peculiar loop already mentioned by Zwierzycki (1919), finally leaves the basin in the NW and runs toward the Indian Ocean.

It is interesting, however, that the Median Graben can be traced on the air photos even within this basin. A number of parallel faults, running more or less in the Sumatra direction, are visible along the Gumpang and Tripa rivers in the southern part of the basin (Fig. 43).

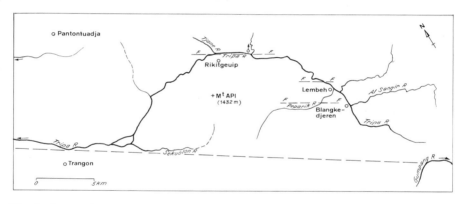

Fig. 43. Sketch of the drainage pattern of the Blangkedjeren basin. Note the great loop of the Tripa River.

Van Bemmelen, in the present author's opinion with good reason, disagrees with Zwierzycki's explanation of the remarkable loop of the Tripa River. The andesites, which according to Zwierzycki blocked the original northwesterly course of the river (by way of the present Peparik and Sekuolon rivers), are non-existent. Mt Api is not composed of volcanic material but consists of Early-Tertiary sediments. Van Bemmelen (1929) and Volz (1899, 1909) assume that a lake once formed in the basin as a result of subsidence along faults. These faults, according to Van Bemmelen, have influenced the drainage. This holds particularly for the fault at least 10 km long running along the Tripa River near Rikitgeuip (Fig. 43).

The present author agrees that the steep gorge formed in the Mt Api ridge by the Tripa River is peculiar, since there is an easy northwestern passage in the Median Graben where the Peparik and Sekuolon rivers now flow. In this zone too, faults were observed from the air. The explanation of the Tripa loop and its transverse gorge in the Api ridge is to be sought in the epigenetic nature of this valley in the graben basin deposits, which, as indicated by the terraces, reached heights of 1,400 m and more.

8. *Deposits and terraces in the Blangkedjeren basin*

High terraces are an outstanding characteristic of the Blangkedjeren basin. They can be seen in the field sketch in Fig. 44, showing four main levels. The two highest levels can be seen in Photo 54, of the western border of the basin near the town of Blangkedjeren. The highest one is situated at an altitude

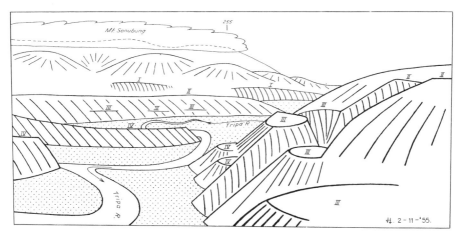

Fig. 44. Sketch of the Blangkedjeren Basin as seen from Mt Lembeh looking WNW. Four main terrace levels can be distinguished on either side of the Tripa River.

Photo 54. View of the western side of the Blangkedjeren Basin, seen from the 1,035 m terrace at Kotapandjang. The two highest terraces are clearly visible.

of 1,350 m, or about 450 m above the river bed. These basin deposits are comparable to the graben deposits mentioned for many other parts of the Median Graben. The highest terrace formed in them indicates the dimensions of the vertical tectonic movements in the past. Such high terraces have not been observed anywhere else in the Median Graben. This fact, and the presence nearby of the lofty Leuser range of the western block mountains, seem to indicate the occurrence of more intense crustal movements here than in most parts of the island.

The heights of the terraces were measured by the author at several localities near Blangkedjeren, Lembeh, and other places in the southern part of the basin. Three high levels were observed, namely at 1,350, 1,225, and 1,100 m, and two lower ones at 950 and 920 m, the latter being 45 and 15–20 m, respectively, above the present river. The 1,350 and 1,225 m levels appear almost horizontal, while the 1,100 m level gradually descends towards the basin via a less distinct 1,000/1,050 m level to 950 m. Blangkedjeren is built on the 950 m terrace, which like the 920 m terrace lies parallel to the gradient of the present river.

The material of the lower terraces is partly clayey and partly composed of sands and fine gravels. Thin beds, rich in plant remains and peat layers, were observed directly to the S of Blangkedjeren. Van Bemmelen (1929) observed similar terraces in other parts of the valley as well. He noted a terrace of 900–1,050 m in the Tjane River near Rikitgeuip and also a 1,050-m high plain, covered by freshwater limestone, on top of a divide near the Tjane River. Along another valley he saw a terrace descending from 1,400 to 950 m.

9. *Northernmost parts of the graben*

The Median Graben NW of this basin is not well known, but the influence of this fault zone on the drainage is pronounced. The fault zone is successively occupied by the upper reaches or tributaries of various rivers, all breaking through the western mountain chain and emptying into the Indian Ocean. A few tiny plains have been formed locally.

Finally, the Atjeh River drains the northwestern extremity of the graben towards Kotaradja. This part of the graben is again very wide, and there are extensive graben deposits in which a few terrace levels can be distinguished, as described by Montagne (1950). A triangular alluvial plain, the base extending along the northern coast of the island, marks the end of this zone in Sumatra. The direction of a series of old beach ridges formed in this plain indicates that some abrasion of the coast has occured to the E and to the W of Kotaradja. The submarine continuation of the Median Graben can be traced in the Bengalen Passage to the W of Weh Island, as is also evident from the 200 m depth contour (see geomorphological map).

d. The mountainous areas to the East of the Median Graben

1. *Sumpur graben*

The dominant feature in this part of the eastern mountains is the steep-sided Sumpur graben (Fig. 38), which becomes wider N of Lubuksikaping. A few rather small alluvial fans occur along the edges of the plain, and the villages, rice fields, and roads are located there. The plain is an area of spontaneous settlement by Batak people emigrating from their densely populated and often severely eroded home area further N. Most of the plain is swampy, however, and is not used for agriculture. It is drained by the NW-flowing Sumpur River, which receives a SE-flowing branch, the Asik River, and finally breaks through the fault scarp bordering the graben to the E and joins the Rokan-kiri River. The narrow transverse valley formed in granites, immediately to the E of the graben, is shown in Photo 44. No terraces occur along this Sumpur River section.

Extensive flat-topped graben deposits can be observed in the Sumpur plain. They are very coarse, and certainly cannot be classed as 'lake deposits'. Occurrences are widespread in the northwestern part of the graben and along the eastern fault scarp (Photo 45).

2. *Asik – UluAer fault zone*

At the northern extremity of the wide Sumpur graben, the fault zone takes a northwesterly direction again and continues in the narrow upper part of the Asik valley.

A remarkable phenomenon is found further on in the Asik fault zone. Over a distance of some 12 km, the fault runs along the eastern foot of the mountains (Fig. 38), where there is an oblong graben (29 x 15) which has no mountains bordering it to the E and which is in open connection with the lowlands of eastern Sumatra. Through this gap, various rivers escape from the Barisan range to the eastern plain. Beyond the gap, the fault again crosses an outward bulge of the eastern mountains in the UluAer valley.*

Another remarkable part of this fault zone occurs approximately in the northwestern extension of the UluAer fault to the E of Padangsidempuan (27 x 15), and can be seen in Photo 55. The Silangkitang and Sababalik rivers run in this fault zone, where the tectonic Lake Tao and the Galinggang swamp are also located. The Siarsiarsik River runs through this zone outside the photograph. Several faults can be clearly distinguished in this area. Photo 55 provides an outstanding example of an extremely young scarplet, locally a rather thin groove resulting from a very recent aftermovement along

---

*Durham (1940) indicates the whole of the Median Graben zone through central and north Sumatra as the UluAer fault zone.

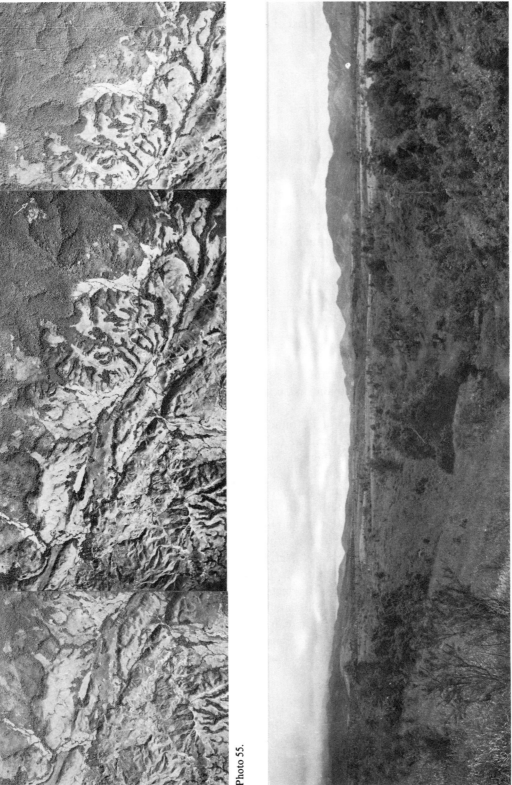

Photo 55.

Photo 56.

Fig. 45. Field sketch of the fluvio-volcanic fan of the Sibualbuali volcano extending eastward. The volcano itself is on the left and the Sipirok graben can be seen in the distance. The mountains in the background to the right are formed by Permo-carboniferous rocks. Looking NW from a point about 2 km E of the Sipirok-Padangsidempuan-Gunungtua road junction.

Photo 55. Stereotriplet showing various geomorphological details related to a fault zone between Sababalikobjai and Sibuhuan, running obliquely from the upper left to the lower right of the centre photo. Note the small elongated Golinggan plain (G) at the extreme left, where a young scarplet (thin groove) can be seen in the alluvium. The small Lake Tao (T) lies slightly further to the W. Stream offsets (O) mark the fault zone in many localities. Scale 1 : 80,000.

Photo 56. The fluvio-volcanic fan of the Sibualbuali volcano, as seen from the Padangsidempuan-Gunungtua road about 2 km E of the junction with the road to Sipirok. The Sibualbuali volcano itself is hidden in the clouds to the left.

the fault. It can even be traced right across the swampy Galinggang plain, near the left edge of the photograph. Numerous stream off-sets can be observed along this fault, but the off-set direction along one end of the fault is just the opposite of that observed near the other edge of the photograph. This example clearly shows how the general slope of the terrain influences the direction of river off-set, and how careful one should therefore be in drawing conclusions regarding horizontal movements along a fault solely on the strength of stream off-sets, particularly in mountainous country.

Slightly NW of the graben just mentioned, a fluvio-volcanic flow of the Sibualbuali volcano spreads eastward in a low zone forming the transition to the Padang Lawas plain, which will be discussed in the chapter on the eastern lowlands. Photo 56 and Fig. 45 show these features, looking NW. Andesitic blocks varying in size are embedded in this lahar material. Some of these boulders have been exhumed by the wide-spread gullying in the

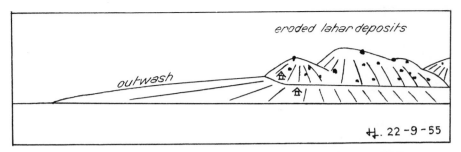

Fig. 46. Field sketch showing the effect of rainwash on the fluvio-volcanic fan of the Sibualbuali volcano about 7 km E of the Sipirok-Padangsidempuan-Gunungtua road junction. Note the barren hill with large lahar blocks at the surface surrounded by the outwash plain, which slopes gently downward.

lahar material, intensified by the scanty vegetation. Exfoliation and splitting of the andesites are common features. Small earth pyramids have formed locally and wash slopes are frequently developed around hills in the lahar products (Fig. 46). Two phases of lahar development can be distinguished in the flow.

3. *Southern part of the Toba plateau*

We now arrive at the extensive Toba plateau (Fig. 47), where the width of the eastern mountain zone suddenly increases from about 15 to more than 70 km. The central part is occupied by the enormous Toba cauldron holding lake Toba. Its eruptions have covered the mountains with ignimbrites, filling up the relief so that only the mountain tops protude from the tuff cover. Along the margins, larger mountain ranges have remained uncovered, and here tongues of ignimbrite/tuffs occupy only the radiating valleys, where they occur as terraces dissected by narrow river valleys.

Various faults can be traced in this southern part of the Toba plateau. The UluAer fault zone, for instance, possibly consinues in the wide Sipirok depression, which is occupied by white ignimbrites. Further N, the Toba ignimbrites form an almost continuous plateau. One of the older mountains emerging from the cover is Mt Paung (24.5 x 17), which Zwierzycki (1919) considered to be an extinct volcano. It is, however, not an easy matter in this region to distinguish between extinct Quaternary volcanoes and Old Andesites.

These (pre-existing) mountains influenced the height of the surrounding ignimbrite cover, which for example is 1,200 m high to the W of Mt Paung and 1,400 m to the E of it. The cover, of course, gradually becomes lower away from Lake Toba, descending in this area from 1,400–1,450 m near the lake to 1,100 S of Mt Paung. This mountain appears to be bordered in the

SW by a fault, although it cannot be traced with certainty due to the Toba tuff cover. The fault movements continued in many localities after the deposition of the Toba ignimbrites. It is quite possible that the zone SW of Mt Paung subsided somewhat in recent times. The mountain is situated along a fault zone, which can be traced to the western edge of Lake Toba.

Fig. 47. Sketch map of the central part of northern Sumatra.

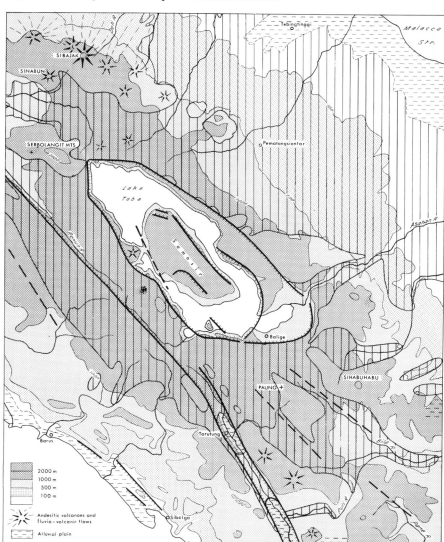

The mountains of the eastern zone of the Barisan can be traced to the E of Lake Toba, where the Asahan River, draining the lake, flows to the Malacca Straits through a breach in this zone. To the N of this valley, the emerged parts of these eastern mountains becomes rapidly narrower until only a few isolated peaks rise from the ignimbrite plateau bordering the northeastern side of the Lake Toba graben. This geomorphologically complex graben deserves special attention.

4. *Lake Toba graben*

The roughly triangular Toba graben measures about 110 km from the NW to the SE, and reaches a width of 31 km at its widest (SE) part. The lake itself is about 87 km long and has a maximum width of 26 km, its level lying at 906 m and the greatest recorded depth being 529 m. The fault scarps bordering the graben are several hundred metres high.

Pre-Toba rocks emerge locally from the ignimbrite cover. The largest occurrence of this kind is the above-mentioned continuation of the eastern mountains stretching along the main eastern Toba fault scarp; a few isolated peaks occur further to the NW, e.g. Mt Simardjarundjung. Along part of the southern edge of the cauldron there is a smaller ridge, along the western fault scarp only one older top, and in the extreme NW the southern tip of the Serbolangit Mountains. From the reminder of the cauldron rim the ignimbrite plateau radiates away unimpeded.

Volcanic activity – although of an entirely different nature – continued after the Katmaian Toba eruptions, as shown by young cones such as Mt Tandukbenua (Piso-piso) located directly to the N of the lake and of MT Pusukbukit (1,982 m), which was formed along the western shores of the lake within the confines of the graben (Fig. 48).

The present author has expressed the opinion elsewhere that the Toba graben existed before the Katmaian eruptions resulted in the formation of the ignimbrite plateau (Verstappen, 1961). He does not share the view held by Van Bemmelen (1939, 1949) that the cauldron was mainly formed after the cataclysm, due to the collapse of the roof of an emptied magma chamber. Since this matter has been dealt with at length in the paper referred to above, the author will confine himself here to a geomorphological description of the complex Toba graben, which according to him belongs essentially to the same category as the numerous other grabens of the island.

Three major morphological units can be distinguished within the graben, namely:
1. a sliver located along the central parts of the main western fault scarp,
2. a tilted block formed by the island of Samosir, which is separated from its counterpart (3) by the narrow Latung Straits, and
3. the Sibolangit or Uluan peninsula.

These features are indicated on the geomorphological map of the Toba Area in Fig. 48.

Fig. 48. Geomorphological map of the Toba graben.
  Key: 1. Ignimbrites of the most recent Toba eruption (arrows indicate places of overflow)
  2. Pre-Toba rocks emerging from the tuff cover
  3. Major fault scarps
  4. Medium-sized fault scarps
  5. Minor fault scarps
  6. Probable fault scarps
  7. Minor faults or grabens
  8. Slivers and other subsided blocks
  9. Post-Toba volcanic cones
  10. Solfataras
  11. Toba terrace I
  12. Toba terrace II
  13. Areas emerging from 1st Toba terrace (on Sibolangit Peninsula)
  14. Multi-cycle fluvio-volcanic fans
  15. Asahan amphitheatre
  16. Post-Toba uplift/warping
  17. Post-Toba tilting
  18. Alluvial plains

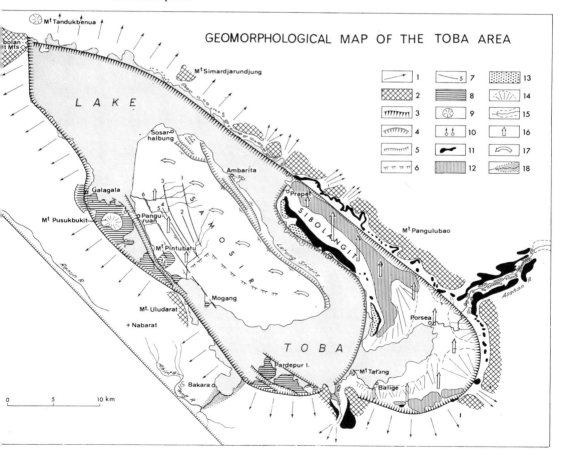

5. *The sliver along the western side of the Toba graben*

The sliver in the W lies in pre-Tertiary rocks but is partly covered by the youngest Toba tuffs and crowned by the young cone of the Pusukbukit volcano (21 x 17). It slopes down from 1,400–1,500 m in the NW to 1,200 m under Mt Pusukbukit. Another very narrow, less important sliver occurs at a greater height near the northwestern extremity of the main sliver. The main western fault scarp shows a westward bend behind this sliver.

A narrow graben, along which hot springs occur, separates the sliver from the island of Samosir to the E. Since a few relics of the sliver also occur along the W coast of Samosir, this narrow graben is evidently younger than the sliver itself, and is not its original boundary. One of the relics is the low flat-topped Pintubatu hill, which is covered by thin layers of volcanic breccias and lake sediments which are, however, very much thinner here than elsewhere. The rivulets draining the lake deposits of the southwestern slopes of Samosir deviate from their consequent direction where they reach this older core and go round it. It is striking to see, even from a distance, how the generally rather gentle slopes in the lake deposits give way to the horizontal surface of this rather isolated plateau. The breccias reported by Westerveld (1947) from Mogang and Cape Tandjungan further to the SE also form low parts of this tectonic sliver zone, slightly above the lake level.

The position of the Pusukbukit volcano on top of the sliver and W of the main zone of youngest Toba deposits indicates that this volcano is younger than the older Toba eruptions, which produced the now solidified ignimbrites, but this does not necessarily mean that it is younger than the younger Toba eruption phases.

Another separate relic of a subsided block within the main Toba graben is found near the southern end of the lake to the E of Bakara. The island of Pardepur, where eastward-dipping, mostly pre-Tertiary beds occur, forms part of it. The relation between this sliver and the one described above cannot be established, since the area between them is part of the lake bottom. The depth contours of the lake (Stehn, 1939), based on a survey by Drost and Bekkering, do not provide any indications.

6. *Diatom deposits of Samosir*

An important feature in the graben is the large island of Samosir, which is considerably younger than the western sliver zone. Its counterpart is formed by the Sibolangit peninsula in the southeastern part of the graben. The island is a SW-tilted block reaching a height of more than 1,600 m near its northeastern edge, where a steep, high, complex fault scarp descends to the lake. The surface of the Samosir Block is formed by its southwestern flank, which slopes much more gently. This is particularly pronounced in the northern part of the island. In the southern half, however, there is a rather sudden transition from the gently southwestward-sloping high plateau to the parts near the coast, which are much lower and deeply dissected. The escarpment separating these two zones in the S cannot be distinguished in the northern half of the island, where a gentle southeasterly slope prevails throughout. The difference in morphology between the southern and northern parts of Samosir has already been mentioned by Wing Easton (1894), but has not been clearly stated since. The same beds (ignimbrites, breccias) cropping out in the eastern fault scarp of the island also occur in this probable fault scarp of South Samosir, along which the southwestern parts of the island presumably slid down.

Samosir is mainly covered by finely stratified whitish lacustrine diatom deposits. Ruttner (1935) found these as high up as 1,360 m. He did not take into consideration the theory of post-eruption uplifting movements in the Samosir area, and therefore came to the conclusion that the lake formerly had a level about 500 m higher than at present and at that time had its outflow in the SE, toward the Toru valley. Since the 1,400 m contour does not close around the lake, he had to assume a regional tilt. Westerveld (1947), who shared Ruttner's view regarding the 1,400 m lake level, tried to solve the difficulty partially by assuming that there was considerable retrogressive erosion in the tuffs. However, this view is geomorphologically unjustifiable, because retrogressive erosion only forms narrow gorges in this material and thus cannot account for the broad gaps in the 1,400 m contour. A former southern outlet of the lake seems to be excluded by the results of a thorough field study by the present author. No dry valley/gorge can be traced here.

The author agrees with Van Bemmelen (1939) that only young tectonic movements can explain the present high location of the lake deposits of Samosir. The highest lake level ever to occur is indicated by the 1,160 m lake terrace (p 125) found in the southeastern portion of the cauldron. The lake has always been drained to the E by the Asahan River.

Van der Marel (1948a) suggested that the higher deposits were formed in small ponds and not in Lake Toba itself. In 1955, however, the present author found an extensive lake deposit in eastern Samosir, at not less than 1,600 m, that certainly was not formed in a small pond. The deposit contained, in addition to sponge needles, rather large amounts of the following diatoms (determination by A. v. d. Werff, Geological Survey, Haarlem):

Melosira distans (Ehr.) Kütz var. robusta Manguin
Melosira granulata (Ehr) Ralfs var. valide Hustedt
Cyclostella stelligera CL. & Grun.

This combination points to a littoral-pelagic freshwater deposit, such as may occur along the shores of large lakes. The present author therefore rejects Van der Marel's hypothesis and assumes that Samosir was almost completely submerged after the youngest Toba eruption and emerged only in more recent times as a result of crustal movements, which caused the general southwestward tilt of the island. The gentle (8–14°) southwestward dip of the lake deposits observed all over the island supports this view. Deviating dips measured locally by the author could all be attributed to slumping, a very common phenomenon in this material, particularly in the S.

7. *Young faults and grabens on Samosir*

A number of very young faults and grabens occur in the youngest deposits of the nothern half of the island. These faults are all approximately parallel, although diverging slightly in the NW, and run in a NW–SE to NNW–SSE direction (see Fig. 48). A detail of one of these grabens (no. 6 in Fig. 48) is depicted in the field-sketch shown in Fig. 49. This area is drained by the small Aron River and, further to the SE, by the Silengga River, both flowing northwestward. Volcanic breccias occur in the lower part of the scarplets, the younger tuffs and lake deposits forming the higher parts. The dips observed at the eastward side of the graben are slightly less acute (9°) than those directly to the W (11°).

The extreme youth of this tiny graben is revealed by the wind gaps or valley 'torsos' occurring in the western scarplet. These gaps indicate that until very recently the right tributaries of the Aron River continued their southwesterly course and reached Lake Toba undisturbed by this graben (Fig. 50). The Padoha rivulet, for instance, was formerly the upper course of the small Djoring River. A terrace (T2) found in this tiny graben and forming the valley bottom slightly further upstream, points to rejuvenation due either to a lowering of the lake level or, more likely, to continued uparching of this part of Samosir.

Interesting too is the course of the Silengga River in the same graben further to the SE, which formed the upper course of the Aron River until very recently, when it was captured by a small rivulet running W of the graben. The fact that the rejuvenation of the river has hardly progressed at all upstream from the capture site, point to the extremely short interval since

Fig. 49. Field sketch of a minor recent graben running NW in western Samosir near Hutantinggi. The graben is drained to the lake by the Aron River.

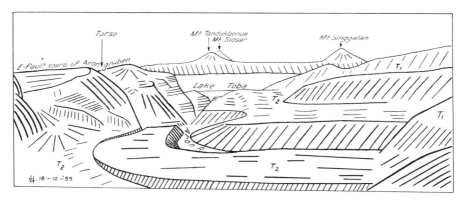

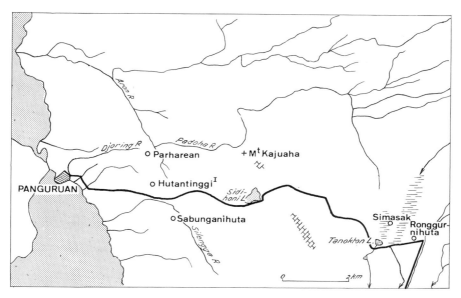

Fig. 50. Sketch of the drainage pattern of part of Samosir to the E of Panguruan, demonstrating the influence of recent fault movements.

this event. Several other wind gaps can also be traced along the fault scarp.

Another interesting detail of this recently faulted area occurs further to the SE, near Ronggurnihuta (Fig. 50). To the N of the road there is a high and shallow valley containing the small Tanokton pond and the small plain of Simasak. Drainage is towards the SW. S of the road the faulted highland ends, the gradient of the river suddenly increases, and there is a steep V-shaped valley descending to the low and well-dissected southwestern part of the island. The impression is gained that this southwestern part of Samosir is a broad subsided zone, bordered by the fault scarp mentioned above.

8. *Fault-scarp morphology of the east coast of Samosir*

The steep NE side of Samosir is also of considerable geomorphological interest. An intensely fractured area is found in the N near the village of Sosorhalbung, where the resistant ingnimbrites cropping out in the upper part of the fault scarp can also be found at several lower levels. All these blocks, which have slid down, are titled eastward, as reflected in the asymmetric valley between the blocks (Photo 57). Volcanic breccias crop out locally, and young tuffs and lacustrine deposits partially cover the structure.

Along the middle part of the fault, there are low coastal hills at the base of the high scarp. The ignimbrites, etc., cropping out in these hills dip gently to the NE. This is particularly clear near Ambarita, where NE dips of 14° and 12–18° were measured 5 km S of the village. A minor graben with

Photo 57. The intensely block-faulted northeastern tip of Samosir, as seen looking S from Sosorhalbung.

paddy fields separates the low hills from the fault scarp in which gentle SW dips occur.

Another interesting and large feature in these parts is the narrow block of hard ignimbrites that reached its present position by sliding down along the narrow fault scarp. The ravines draining the fault scarp have carved V-shaped cuts into this subsided 'slice', downstream from which coarse fans have formed.

The whole structure, which may well be classed as a faulted flexure, is clearly visible in Fig. 51 and Photo 58. The fact that even the lacustrine deposits have been affected by the above-mentioned northeastward dip means that the structure came into being, or at least underwent its main development, after the most recent Toba eruption. This is perfectly consistent with

the observations of SW-titled lake deposits (p 128) elsewhere on the island, and it points to the emergence of Samosir after the last eruptions. This might also explain why pumice, so abundant in other parts of the Toba area, is almost completely absent on the island, and why lacustrine deposits occur even in its highest parts. It also offers an adequate explanation of the absence of lake terraces.

Photo 58. View of Lake Toba, looking SE from the eastern shore of the island of Samosir, N of Ambarita. The flexure in this part of Samosir is clearly visible: the main escarpment, where the ignimbrites dip gently toward the SE, appears at the right (1); directly in front of it there is a detached sliver (2). The ignimbrites dip in the opposite direction in the low hills (3) forming the peninsula and the islet in the centre. The Sibolangit peninsula can be seen in the distance in the centre (4) and the main western fault scarp of the Toba cauldron is seen at the left (5).

Fig. 51. Field sketch of the northern part of the fault scarp (flexure) bordering Samosir to the E. Note the detached sliver (S) in front of it and the fan formed in a breach of the sliver.

The flexure becomes steeper and is finally replaced by a simple but huge fault scarp further to the SE where the Sibolangit peninsula is located below the main eastern fault scarp of the Toba cauldron, separated from Samosir only by the narrow Latung Straits.

9. *Fault-scarp morphology of the Sibolangit peninsula*
The northern part of the fault scarp bordering the Sibolangit peninsula to the W runs in a SE direction, but turns S and even SW as it approaches Porsea Bay, which in the present author's opinion should be regarded as the drowned upper part of the Asahan valley. Since the lake terraces (p 138) do not disappear below the lake level near Porsea, it is evident that this bay existed prior to the last eruption (± 2000 years? v. Bemmelen, 1949). It is older even than the earliest Toba eruption, according to the present author, as indicated by the presence of old Toba ignimbrites (< 300,000 years; Katili, 1969) in this valley. The fault scarp can also be traced at the other side of Porsea Bay, along the western edge of the Mt Talang peninsula to the W of Balige (Fig. 48), where it is 1,070 m high. Photo 59 is a vertical aerial view of this part of the fault scarp. The large concave landslide scars give it a somewhat irregular outline.

Recent faults, like those reported for Samosir, are also found in the young whitish Toba tuffs of the Sibolangit peninsula and caused, for example, the straight courses of a few rivulets in the southern extremity near Porsea. A particularly interesting example was observed on the air photos of an area between Prapat and Porsea and is indicated in Fig. 52 (scale 1 : 100,000). Two pairs of young scarplets cross here at almost right angles, slightly to the S and at the foot of the main eastern Toba fault scarp. Many other similar

Photo 59. Vertical aerial view of the peninsula E of Balige, forming the continuation of the Sibolangit peninsula. Large concave landslide scars occur along the fault scarp, from which the terrain slopes gradually down toward the SE. The stratification of the material can be clearly seen in the scars, and the beds dip approximately parallel to the terrain itself. Scale about 1 : 20,0000.

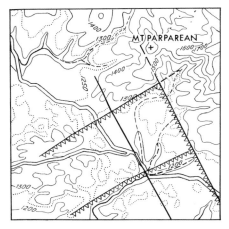

Fig. 52. Schematic map showing young scarplets resulting from recent fault movements in the Toba ignimbrites on the Sibolangit peninsula near the Prapat-Porsea road S of Mt Parparea. Scale 1 : 100,000.

Photo 60. Vertical aerial view of a severely eroded part of the youngest Toba ignimbrites NW of Porsea. Scale about 1 : 15,000.

Photo 61. Ground view showing details of the erosional forms and steep slopes maintained in Toba ignimbrites (cf. Photo 60).

young scarplets may exist, but their detection is not easy because the air-photo coverage is very incomplete.

It is noteworthy that the young unconsolidated Toba tuffs are rather susceptible to erosion. Steep, almost vertical, slopes characterize these erosion gullies, and fissures are also found. Photos 60 and 61 give an aerial and ground view, respectively, of these features near Porsea. The unconsolidated tuffs have a high permeability and are partially drained by a subterranean network of tunnels. Occasionally, collapse of parts of the tunnel roofs leads to sudden subsidence and the formation of small closed depressions at the surface.

10. *Lake terraces on the Sibolangit peninsula and along the southern lake shore*

Extensive lake terraces occur, particularly in the Sibolangit peninsula and also along the southeastern part of the Toba cauldron. Their occurrence has been mentioned by several authors, but they had not been studied in detail before the present author visited the area. They are not present everywhere

Photo 62. The upstream end of the Proto-Asahan valley, looking E from a point near Porsea. The valley was filled by Toba ignimbrites, which led to the formation of the clearly visible 1,160 m lake terrace (I). The roughly 970 m high Toba terrace (II) can be seen in the foreground. The present Asahan River, which has cut a deep gorge in the ignimbrites, flows from the right to the centre of the picture.

around the lake, because where hard rocks occur the surf action in the lake was not sufficiently vigorous to form terraces, and furthermore because slumping – a common phenomenon in the diatom deposits – destroyed the terraces in other localities. Recent emergence due to young tectonic movements is another cause of the absence of terrace remnants, as is the case on Samosir. The best-preserved terraces are found in the extensive tuff deposits in the southeastern part of the graben. Their study is complicated by the frequent occurrence of sub-horizontal rocks, where there has been formation of structural (pseudo) terraces or rock benches lacking any signifcance for

Fig. 53. Field sketch of the upstream end of the Proto-Asahan valley, looking E from a point near Porsea. The 1,160 m lake terrace is clearly visible; the 970 m terrace appears in the foreground (*cf.* Photo 62).

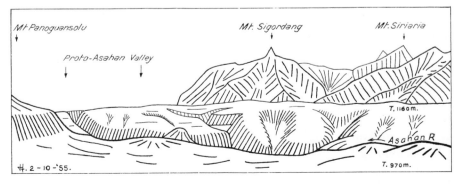

the purpose of establishing former lake levels. Many older observations must therefore be rejected as incorrect.

Two main terrace levels were distinguished. The highest one is represented by the 1,160 m level at the entrance to the Asahan valley near Porsea, visible, in Photo 62 and Fig. 53. This tuff terrace corresponds to the upper level of the tuff terrace found along this river, as mentioned on p 141. This terrace (I) was observed almost without interruption along the eastern fault scarp S of Prapat, and some scattered remains were found along the southern side of the graben E of Balige. It is also developed on the Sibolangit, peninsula, where only a few small areas emerge from it, representing former islands (Fig. 48). A second important terrace (II) is best developed along the Prapat-Porsea road, between the remnants of terrace I on the peninsula and the eastern fault scarp. Its height near Porsea is 1,050 m.

The strong tilt of these terraces can easily be seen in the field, even with the naked eye (Photo 63). Important post-interruption tectonic movements,

Photo 63. Distinctly tilted part of the 1,160 m terrace (I) at the foot of the eastern fault scarp of the Toba cauldron SE of Prapat. (*cf.* the diagram in Fig. 54).

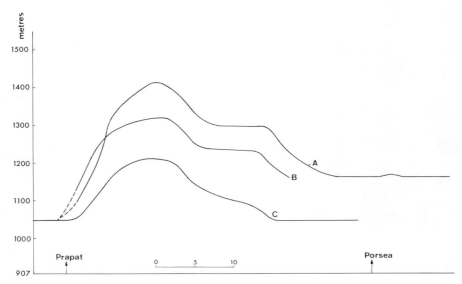

Fig. 54. Diagram of the warped lake terraces along the eastern shore of Lake Toba between Prapat and Porsea.
  A. Terrace I, along the eastern fault scarp
  B. Terrace I, on the Sibolangit peninsula
  C. Terrace II, along the Prapat-Porsea road (between A and B)

resulting in terrace warping, occurred between Prapat and Porsea, as indicated in Fig. 54. The height of the 1,160 m terrace varies from 1,070 m to 1,420 m. It is possible that this terrace – and thus the maximum lake level – originally lay about 1,050 m above sea-level and that the Porsea section was later uplifted some 110 m, thus forcing the Asahan River to increased incision. No warping of these two main terraces seems to have occurred along the southeastern escarpment of the Toba cauldron. The extensive tuff deposits in the southeastern part of the Toba graben make it evident that the more recent Toba eruptions were centerd there (van der Marel, 1947, 1948).

Small remnants of a number of still lower terraces also occur, having their counterpart in the multi-cycle fans formed in the lower parts of the tuff deposits. Photo 64, of a part of the southern escarpment between Porsea and Balige, shows, for example, a well-developed relic of a terrace at 1,010 m. S of Balige, these terraces slope down toward the lake (Fig. 55) and then rise to great heights further upstream. No assumption of tilting movement is necessary to account for this, since here this probably concerns huge multi-cycle fans, formed where a (structural) valley enters the cauldron.

Small relics of lower terraces can also be observed all along the northeastern and southern shores of the lake. E of Porsea, levels were observed at 913, 930, 950 and 1,000 m. Of these, the 930, 950 and 1,000 m levels could

Photo 64. The escarpment bordering Lake Toba to the S between Porsea and Balige, as seen looking toward Balige. A low terrace (at 1,010 m) is well developed near the village of Simarmar.

also be traced NW of Prapat near Hutamanik. Warping of these terraces was not observed; if present, it must be much less marked than that observed in the major terrace levels occurring at greater heights. It should be noted that E of Porsea there are also relics of terraces at 1,100 and 1,130 m, but these were not traced elsewhere. The lower terraces usually take the shape of multi-cycle fans where unconsolidated Toba tuffs and lacustrine deposits occur and particularly where the steep main escarpment of the Toba graben

Fig. 55. Field sketch of the terraces sloping down to the lake, S of Balige. These features may be called multi-cycle fans.

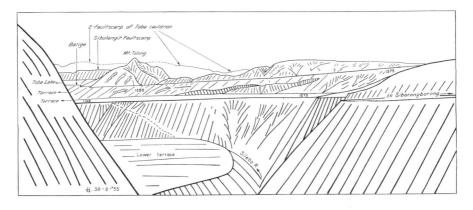

Fig. 56. Field sketch of one of the multi-cycle fans surrounding the southern escarpment of the Toba cauldron. Looking SW from the Hutasalim leper colony.

is nearby. Rainwash and fluvial reworking of these materials locally has produced distinct crossbedding. Fig. 56 shows an example of such multi-cycle fans, the various levels of which are related to the step-wise lowering of the lake level.

11. *Asahan valley*

It is important that the Toba lake terraces can be correlated with structural terraces (Fig. 57) formed in the Toba tuffs/ignimbrites along the Asahan valley. Photo 65 shows the structural Asahan terraces as seen from a point near Siguragura Falls. Fig. 58 gives another view of these phenomena. The intermittent lowering of the lake level is readily explained by the different rates of incision of the Asahan River in the various volcanic strata. Every hard bed in the valley corresponds to a lake terrace. The 1,160 m ter-

Photo 65. Structural terraces in the Toba ignimbrites filling the Asahan valley. The mountains in the distance consist of Tertiary rocks. Seen looking SE from a point near the **Siguragura falls**.

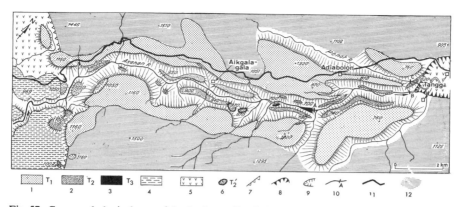

Fig. 57. Geomorphological map of the Asahan valley draining Lake Toba.
Key: 1. Structural terrace I, corresponding with 1,160 m lake level
2. Structural terrace II, corresponding with 1,050 m lake level
3. Structural terrace III, associated with Siguragura falls
4. Toba lake terrace at 970–950 m
5. Alluvial deposits
6. Structural terrace T II' (870 m)
7. Main fault scarp of the Toba cauldron
8. Escarpments of structural terraces
9. Escarpment of Asahan amphitheatre
10. Waterfalls: A. Siguragura; B. Sampuran Harimau; C. Ponot.
11. Road
12. Pre-Tertiary, Palaeogene, and granitic rocks bordering on Asahan valley

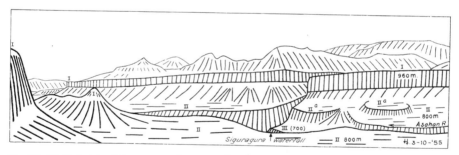

Fig. 58. Field sketch of the structural terraces developed in the Toba ignimbrites in the Asahan valley. Three main terraces can be distinguished, looking E. Terraces I and II are the equivalent of the two main terraces around Lake Toba. The Siguragura waterfall is located directly to the left of the lowest terrace III (700 m). See also Photo 65. The mountains in the background consist of Tertiary rocks.

race is very wide, and can be traced all along the Asahan valley. It is formed in the unconsolidated youngest white Toba tuffs/ignimbrites containing abundant pumice, and it descends to about 950 m near Siguragura Falls and to 750 m near Sampuran Harimau Falls at Tangga, about 15 km downstream from the Toba cauldron. A brick-red tuff layer with numerous

141

Photo 66. The 100 m high Siguragura falls of the Asahan River, with a deep narrow gorge of a minor branch river. Note the extremely steep slopes in the ignimbrites.

quartz grains occurs in a slightly lower terrace having a height of 1,130 m in the upstream end of the valley and about 925 m at a point 9 km further downstream. Lower structural terraces have formed in the older, solidified ignimbrites, e.g. the 800–850 m terrace found between Siguragura Falls and Aekgalagala, located at an altitude of 750 m near Adiabolon. This bed represents the upper limit of the hard ignimbrites and corresponds to the 1,050 m lake terrace (II).

Immediately above the 100 m high Siguragura Falls, another structural terrace occurs at approximately 700 m, and 3 km upstream reaches a height of 870 m. The Siguragura and the Sampuran Harimau falls formed in two resistant and still lower ignimbrite beds. Two more lake terraces will be formed when these falls have reached the lake as a result of retrogressive erosion. Photos 66 and 67 give an aerial and a ground view, respectively, of the 100 m high Siguragura Falls. The large amphitheatre formed near Tangga at the eastern end of the ignimbrite area is shown in Photo 68. Here, the Asahan River passes over the 150 m high Sampuran Harimau Falls and the small Ponot River drops a sheer 280 m slightly further downstream;

Photo 67. Stereopair of the same area as Photo 66. Note the high structural terrace on either side of the Asahan River, and the lower one near the bridge, where the construction of a power plant has been proposed. It is noteworthy that the straight Asahan gorge is considerably narrower than that of its small branch, due to the more resistant ignimbrite bed. Scale about 1 : 20,000.

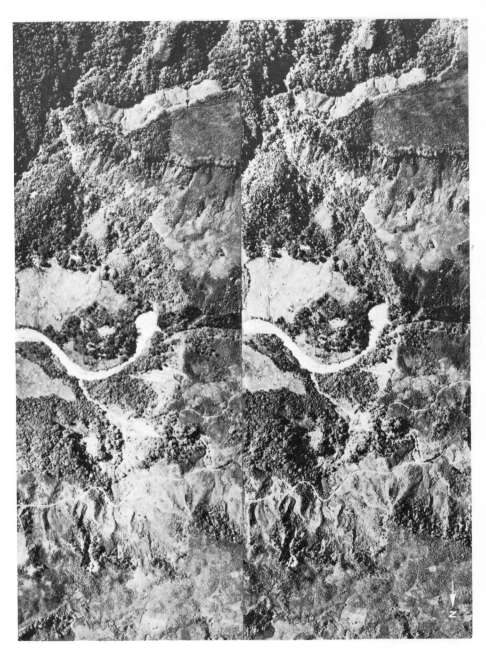

Photo 68. Stereopair of the amphitheatre near Tangga in the Asahan valley at the downstream end of the Toba ignimbrites with which it is filled. The two main structural terraces formed in the ignimbrites, which are responsible for the two main lake terraces of Lake Toba, can clearly be seen. The Asahan River here forms the Sampuran Harimau falls, and its small branch, the Ponot River, has an even higher fall. Scale about 1 : 20,000.

Photo 69. Ground view of the 150 m high Sampuran-Harimau falls of the Asahan River near Tangga. Compare with Photo 68.

Photo 70. Ground view of the 280 m high fall of the small Ponot River in the Asahan amphitheatre near Tangga. Compare with Photo 68.

ground views of these two falls are given in Photos 69 and 70, respectively. A geological study of the Asahan valley and surroundings was carried out by Frylingh (1925). A set of vertical air photos and the detailed contour maps made from them by Djawatan Fotogrametri Central, Bandung, greatly facilitated the geomorphological study of the valley.

The stratification of these older Toba products indicates that a large number of eruptions occurred, rather than the one or two assumed previously by Van der Marel. Parallel slope recession is characteristic of this material and has led to the formation of a true canyon.

The Asahan Riveer is deeply incised in a narrow gorge, locally slightly broader but everywhere showing extremely steep or vertical sides. The broader stretches may be the result of the sliding down of thin ignimbrite 'slices' along the steep slope after undercutting by the river. Van Bemmelen (1929b) mentions that around 1914 a slump occurred near the upper end of the valley, causing the level of Lake Toba to rise approximately 1.5 m. This obstruction had to be blown up, because it had led to the flooding of many paddy fields around the lake.

12. *Karo Highlands and surroundings*

The Karo Highlands, formed on the Toba tuffs NNW of the lake, deserve further discussion. They are bordered on the N by a row of volcanoes and a mountain chain formerly known as the Van Heutsz Mountains (18 x 18) and on the W by the Serbolangit Mountains. Erosion and piping (suffosion) have caused a peculiar, irregular surface in parts of these highlands (Photo 71). A thin layer of andesitic material derived from the volcanoes to the NW makes these parts more fertile than the ignimbrite plateau elsewhere in the Toba Area.

The Karo Highlands are drained by two parallel rivers, both running to the W. One is the Biang-Wampu River, which turns N, crosses a range formerly called the Van Heutsz Mountains, and debouches into the Malacca Straits. Toba tuffs occur in the valley and form a narrow strip further downstream in the eastern plain of Sumatra. A faint ridge in the highlands forms the main divide of Sumatra and separates this drainage basin from the Kitelangkah River, which turns S, crosses the Serbolangit Mountains, draining toward the low Renun Section of the Median Graben, and finally ends

Photo 71. The Karo Highlands with the Sinabung volcano near the southern end of the Serbolangit Mountains, looking toward the E. Note the peculiar topography developed in the ignimbrites.

Fig. 59. Field sketch of the 590 m terrace (T) formed in the Toba ignimbrites in the valleys of the Lisang and Perimboon rivers between the Karo Highlands and the Renun graben. The Gunung River, which has a steep gradient and drains to the Renun section of the Median Graben, captured these rivers.

in the Indian Ocean. Young Toba tuffs can also be distinguished along the Kitelangkah River, where tuff terraces are well developed.

It may be assumed, in view of the terrain configuration, that this drainage came into being when the original drainage to Lake Toba became blocked by the Toba ignimbrites. The narrowness of the gorges indicates how recently these rivers broke through the mountains and established their present courses. This is particularly clear in the case of the Kitelangkah River. The tuff terraces can be observed all along the river and its tributaries. Fig. 59 shows the situation slightly to the E of the Renun graben, where the terrace has a height of 590 m.

The Gunung River, a short Renun tributary with a steep gradient, captured the Kitelangkah River drainage basin after the Toba eruptions, as a result of strong retrogressive erosion probably related to the further subsidence of the Renun graben, which at present has an altitude of only about 200 m. Photos 72 and 73 show the terrace in the Kitelangkah and tributary valleys.

Photo 72. Terrace formed in the Toba ignimbrites in the Lisang valley at an altitude of about 600 m; looking upstream toward the NW.

Photo 73. The same terrace as in Photo 72 but shown here along the Kitelangkah River as seen from the Lisang capture. Looking upstream toward the SE.

The 590 m terrace can still be clearly seen along part of the steep Gunung valley. Its height above the river increases rapidly after the river descends to the Renun graben. The exact location of the capture can be seen from the abrupt termination of the 590 m terrace.

13. *Stratovolcanoes bordering on the Karo Highlands*

Along the northeastern border of the Toba plateau, a row of old volcanic mountains rises from the ignimbrite plateau, parallel to and at a distance of about 25 km from the lake. Towards the NW, young, post-Toba stratovolcanoes occur instead. These phenomena seem to be accounted for by a fault scarp running in the Sumatra direction, along which volcanic activity gradually shifted northwestward. The Sibajak (2,094 m) is the only active volcano of the group; solfatara occur at its top and also on its northern and southern slopes. The fault referred to above can be traced along part of the W foot of these volcanoes.

The Sinabung volcano takes a more isolated position in the northern part of the Karo Highlands (19 x 18). The cone of this volcano shows a V-shaped cut to the S somewhat below the top, where solfatara occur. Several old lava flows can still be traced on its slopes. One of these can be seen as an unvegetated flow on the northern slope, below about 2,100 m. Slightly further to the E, an older flow descended in several branches and dams a river valley, thus forming Lao Kawar (*lao* = lake). Other flows took various directions, and a complex flow was noticed on the SSW slope.

The presence of signs of recent volcanic activity accounts for the fertility of the Karo Highlands, but it is more important that their extensive fluvio-volcanic products spread far into the low eastern zone, to the benefit of the flourishing estate agriculture (tobacco, etc.) of the Deli district and the growth of the town of Medan.

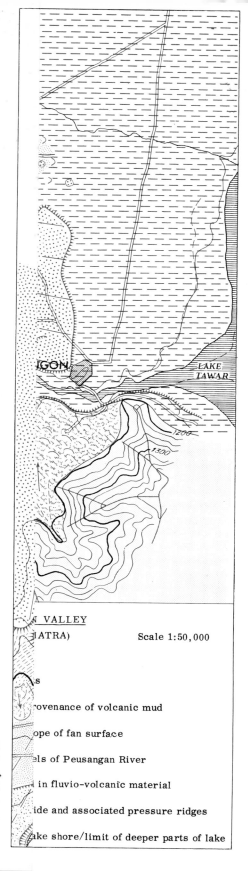

Fig. 62.
Preliminary geomorphological map of the Peusangan valley downstream from lake Tawar, Atjeh. Fluvio-volcanic material coming repeatedly from the NW blocked the river, which subsequently formed terraces here. Scale 1 : 50,000.

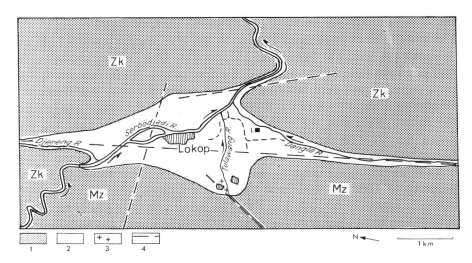

Fig. 61. Sketch of the Lokop graben, northern Sumatra, drawn from an aerial photograph. Scale about 1 : 70,000. Several faults can be recognized in this area, which is drained by the Serbeudjadi River. Hot springs occur near the Taluwang River, which is controlled by a fault. An old course of the Dangla River can clearly be seeen near the village of Leles (L).
  Key: 1. Mountainous areas surrounding the graben
        Zk = Black claystones (Upper-Palaeogene)
        Mz = Micaceous sandstones, underlying Zk
     2. Plain of the Lokop graben
     3. Hot springs
     4. Faults

with a fault clearly visible from the air. The lithology indicated in Fig. 59 is based on observations made by 't Hoen (1929), who compiled the geological map Atjeh II, 1 : 100,000.

15. *Gajo Districts*

An interesting study of the Gajo Districts in the N was made by van Bemmelen (1930). The paper by Oppenoorth and Zwierzycki (1918) and books by Volz (1899, 1909) also provide information concerning these poorly known parts of Sumatra. In the present discussion of the Blangkedjeren basin or Gajo Loeës (p 116), it has already been mentioned that the northernmost parts of the Barisan Mountains are characterized by a number of broad basins or depressions, which include, besides that of Blangkedjeren in the Median Graben, the basins of Gajo Deureut and Gajo Serbeudjadi in the eastern mountains. The small Lokop graben located just E of the mountains, forms part of the Serbeudjadi basin. Considerably larger is the Gajo Deureut basin (11 x 18), which is drained by the Djaboaje River.

Oppenoorth and Zwierzycki (1918) claim that two erosion cycles can be distinguished in these areas, the oldest one having reached an advanced stage of maturity when renewed uplift initiated the second cycle. These authors place the beginning of the first cycle at the end of the Palaeogene and the subsequent uplift of about 900 m at the end of the Neogene. This opinion is not, however, shared by Rutten (1927) or by Van Bemmelen (1929) who maintain that the first cycle did not start until the Upper Neogene. This would account for the fact that when the second cycle started only an advanced maturity had been reached instead of complete bevelling of the mountains. The dissection of the terraces observed by the present author in the Blangkedjeren basin, most of which were formed in Palaeogene materials, is consistent with this. The same terraces were also observed by Van Bemmelen in other parts of the same basin, and are accounted for by the second uplift. Similar phenomena can be expected in other parts of the Gajo Districts.

16. *Laut Tawar and the Peusangan fluvio-volcanic terraces*

An outstanding feature to the W of the Gajo Deureut basin is a slightly curved graben, striking generally E–W, at the westernmost end of which the large Lake Tawar is located. The divide between the Gajo Deureut basin and the Peusangan drainage basin is located in the graben, approximately 5 km E of the lake. The graben is bordered on the N and S by high ridges rising up to about 2,500 m. Three minor extinct volcanic cones occur W of the lake.

Much more interesting is the active Bur ni Telong stratovolcano (2,600 m), to the N of which the extinct cone of Mt Geureudong rises to about the same height (2,590 m). Photo 74 gives a vertical aerial view of the Bur ni Telong volcano with its crumbling lava plug. Another but smaller plug is located on the southwestern Telong slope. The fluvio-volcanic fans of this volcanic group spread far northward, covering extensive parts of the lower coastal hills. The Peusangan River loops round the southern and western edges of this volcanic region, and extensive terraces are found in the volcanic products in its middle and lower valley.

A set of air photos on a scale of 1 : 20,000 revealed the existence of important fluvio-volcanic terraces in the Peusangan valley immediately downstream from the outlet of the lake. A photogrammetric map compiled from these photos at ITC, Eschede*, served as the basis for the preliminary geomorphological map of the area (Fig. 62) based on photo interpretation. No past volcanic activity of the Katmaian or any other type can be traced on the photos of the Tawar graben. However, at least two major fluvio-volcanic flows (I and II) coming from the NW once blocked the Peusangan River a few kilometres downstream from the outlet of the lake.

It can be seen from the map (Fig. 62) that these fan-shaped deposits slope

---

* The author is indebted for this map to J. Visser, photogrammetric engineer at ITC.

Photo 74. Vertical aerial view of the active Bur ni Telong volcano (2,600 m), located N of Lake Tawar. Note the irregularly shaped and crumbling lava plug forming the top, from which fumes emerge. Several explosion craters (C) can be seen in the plug. Note also that the separate plug (P) of Mt Pupok (1,681 m) in the SW corner and the SW-running barranco in the top, are drained by the Latas River (L). The upper reaches of the Djernih (D) and Lampahan (Ln) rivers drain the western slopes of the volcano. Average scale 1 : 13,000.

down, both toward the lake in the E and downstream toward the S. Evidently, a temporary rise of the lake level resulted from this blocking of its outlet, after which the lake waters found a number of new outlets across the fluvio-volcanic material. Finally, only one of these, the southernmost, became the permanent drain of the lake. This process was repeated at least twice, and the related fans, outlets, and other phenomena are indicated on the geomorphological map with different symbols. The Peusangan River, coming from the lake, has of course only a negligible load, and must have started incising its course directly after the volcanic events.

Younger, lower terraces are found in the fluvio-volcanic material and cover a large area shown in the southern part of the map. All these terraces slope distinctly toward the SE, in other words away from the vertex of the fluvio-volcanic fan structure in the NW. Therefore, they are not river terraces in the true sense. They must be the result of either repeated fluvio-volcanic activity or, more likely, intense washing out of fan II (and I?) during the incision of the Peusangan River. True river terraces and a few small recent alluvial cones of minor tributaries occur only at much lower levels above the river. A detailed investigation of this unique phenomenon of repeated fluvio-volcanic damming of a river valley would be of considerable geomorphological interest.

17. *Northernmost areas and the island of Weh*

Little can be said regarding the eastern mountains further to the NW, where this zone becomes increasingly narrow W of the Peusangan valley. Running parallel with lake Tawar there is also a longitudinal valley with another, smaller lake. This latter valley may belong to the same graben zone; it ends at the isolated, active, Peuet Sagoëe volcano (2,801 m). An E–W-striking row of craters occurs in its top area. The oldest one is in the W, and slightly to the SE of it is the active crater with solfatara activity. To the E is another extinct crater, and the top of the volcano is found in the easternmost parts. Two extinct cones rise in the extreme northern part of the eastern mountain zone. The Seulawaih Agam volcano (Goudberg; 3 x 17.3; 1,762 m), whose products spread freely toward the Median Graben and also to the NE and E, is the larger of the two. These volcanoes also cover parts of the folded Tertiary sediments forming the hills and low ridges near the coast.

Off the northern coast of Sumatra, the island of Weh forms the continuation of the eastern mountains. This island is mainly composed of old volcanic rocks, but it appears to the author that both the western peninsula and a belt along the eastern coast are tilted blocks between which there is a lower graben zone to which Sabang Bay belongs. Young volcanism occurs in the S, where an active crater with solfatara is found. Zwierzycki (1916) mentions widespread terraces on the island at 20, 40, and 100 m altitudes. The town of Sabang is built on the lowest of them, and the 40 m terrace can be traced all around Sabang Bay. The 100 m terrace climbs to 160 m in the S. Remains of coral reefs are found on these terraces. A considerable uplift of the island is evident from these terraces, which are correlated with those mentioned from many parts of northern Sumatra.

e. The low areas of eastern Sumatra

1. *Padang Lawas and the plain E of the Toba ignimbrites*
The low eastern zone, consisting of folded Tertiary sediments, and the coastal plain, is rather narrow throughout northern Sumatra, and it might have been better to treat it in conjunction with the eastern mountains. It is treated separately merely for the sake of consistency with the discussion of central and southern Sumatra.

The southernmost part is formed by the Padang Lawas plain (27 x 16), extending at the southeastern foot of the Toba plateau. It is bounded in the E by the volcanic mud flows of the Sibualbuali volcano (see p 121) and by the low eastern Barisan ridge consisting of Permo-carboniferous rocks. The area is known for its comparative dryness and particularly for the föhn winds which blow frequently (p 11). The morphological characteristics of the low parts of the Barisan range just described are the cause of these climatic peculiarities. The scanty vegetation in its turn favours strong rainwash and erosion.

Along some main rivers a broad 7–9 m terrace is found and another at a level of about 20 m. Agriculture is concentrated on these terraces, above which stretch the dry eroded plains. These shrink to a rather narrow zone further to the NW, and finally disappear below the lower portions of the Toba ignimbrite sheet. The flows of Toba ignimbrites extend far into the coastal plain, and only a narrow alluvial plain separates them from the Malacca Straits. The series of beach ridges found in this plain are considerably more numerous and broader than further to the S, which is easily explained by the proximity of the hinterland. Deep steep-sided valleys formed in the ignimbrites as a result of the weaker gradient of the river as compared to the original slope of the ignimbrite sheet. Therefore, the volcanic products must have yielded sufficient amounts of sand and silt for the formation of the beach ridges.

2. *Alluvial plains in the North and their hinterland*
It is only to the NW of the ignimbrites and the andesitic lahars forming the hinterland of Medan that Tertiary rocks reappear. The Palaeogene rocks make up the low mountains, whereas in the Neogene areas only low (50–150 m) hills occur. It was observed by 't Hoen (1919) that the folding of the Atjeh parts of the eastern geosyncline often is most intense at the western or Barisan side; more gentle folding occurs in the E. An interesting geomorpho-

Photo 75. Vertical aerial view of the finely striped beach ridges in the alluvial plain near the Pendawa Raja River, Atjeh. Scale 1 : 14,000.

logical aspect is produced by the resistant Keutapang sandstone horizon, which often forms high ridges (see geomorphological map) and frequently results in a complete inversion of the relief.

At the seaward side, Quaternary flat-topped marine terraces, composed of horizontally bedded unconsolidated gravels, sands, and sometimes clays, are reported. Their extent is not completely known and the geomorphological map is certainly inaccurate for this region. These terraces have their counterparts in river terraces further upstream in the valleys. The monotony of the narrow alluvial plain is interrupted by numerous beach ridges, Neogene hills emerging locally, and these terraces. Estuaries are a common feature along this part of the east coast, explained by Weeda (Baartmans *et al.*, 1947) as due to the rather large tidal range prevailling here.

Photos 75, 76, and 77 are vertical aerial views of the beach ridges in the alluvial plain of eastern Atjeh. Photo 75 depicts a finely 'striped' beach-ridge belt near a river mouth. Photo 76 shows an intricate beach ridge complex,

Photo 76. Vertical aerial view of the beach ridges in the alluvial plain near Cape Intem, Atjeh. Three subsequent sets of beach ridges can be distinguished, each one cut off by the next, thus demonstrating changes in the directions of the coastline during the development of the plain. Scale about 1 : 24,000.

Photo 77. Vertical aerial view of part of the alluvial plain bordering the northern coast of Sumatra at the mouth of the Teupinan River. Numerous sandy beach ridges covered with coconut palms are separated by shallow moats, some in use as fish ponds but most of them drowned by an incursion of the sea. Abrasion of the coast is demonstrated by the beach ridges cut off by the present coastline at the lower left. Scale about 1 : 24,000.

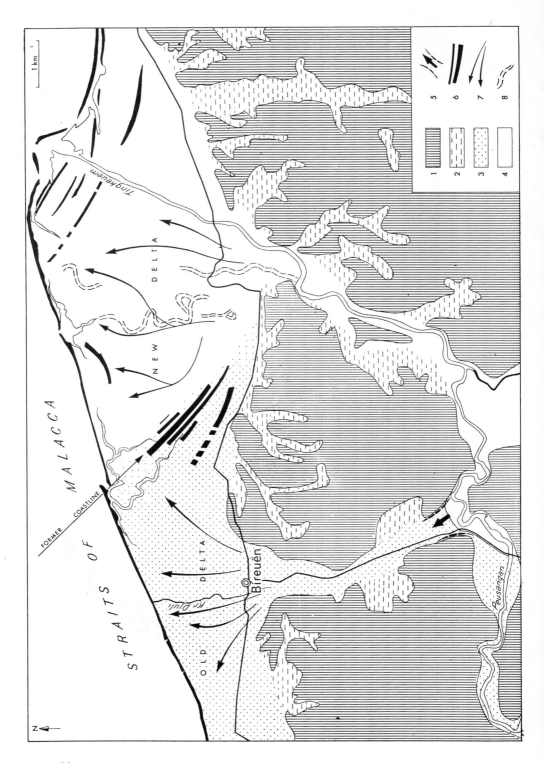

where three successive sets of ridges intersect one another, indicating important changes in the coastline. Photo 77 gives a picture of part of Atjeh's northern coast, where ridges are cut off by the present coastline, indicating local abrasion.

A particularly interesting change in the coastline observed near Bireuen is the result of a change in the lower course of the Peusangan River. A few thousand years ago this river must have debouched near the present town of Bireuen, but it was captured, about 8 km S of the town, and now enters the sea about 16 km further to the E. The old course and the capture site form a broad terrace about 20 m above the present river. The old delta is severely abraded, now showing a rather straight coastline, whereas the new delta to the W is still growing. Fig. 63 shows the situation, which has been dealt with in more detail elsewhere (Verstappen, 1964).

The area occupied by the eastern belt of folded Tertiary sediments gradually becomes lower and disappears below sea-level, as described above (p 94), and consequently the island of Sumatra becomes narrower toward its northwestern tip.

Fig. 63. The effect of river piracy on the development of the Peusangan delta (northern Sumatra)
    Key: 1. Hilly hinterland (Neogene; E and W of Bireuen: Pleistocene terraces)
        2. Minor valleys in the hinterland
        3. Delta formed prior to the capture
        4. Same, after the capture
        5. Site of the capture
        6. Beach ridges indicating former and present coastlines
        7. Approximate location of former Peusangan courses (natural levees)
        8. Distinct recently abandoned Peusangan channels

VIII The Island Festoon to the west of Sumatra

a. Introduction

This archipelago which, reading from S to N, comprises the island of Enggano, the Mega reef, the Mentawei Islands (i.e. North and South Pagai, Sipora, and Siberut), Tanahmasa and Tanahbala of the Batu group, and the islands of Nias and Simalur, represents the non-volcanic arc accompanying the Sumatra section of the volcanic arc. Its greatest height (886 m) is reached on the largest island, Nias.

The arc is separated from Sumatra by an interdeep having a maximum depth of 2,364 m and interrupted by three transverse thresholds tying the non-volcanic arc to the volcanic island of Sumatra. The southernmost of these is very inconspicuous, even completely submarine, and is located S of the Mentawei group and NE of the Mega reef. The second lies within the Batu group of islands, and is clearly indicated by the 200 m isobath (see map), by the island of Pini, and by the seaward bulge of the Sumatran coast between Airbangis and Natal. The northernmost transverse threshold is represented by the Banjak archipelago, and is also indicated by the 200 m isobath and the seaward bulge, near Singkel, of the Sumatran coast. N of the Banjak threshold, the volcanic arc, the interdeep, and the non-volcanic arc (Simalur section) curve more strongly toward the NW.

W of the non-volcanic arc, the foredeep – showing strong negative gravity anomalies – can be traced northward to Simalur as a trough reaching depths of 5,000 to 6,000 m.

Literature concerning the developmental aspects of the land forms of the island festoon is rather scarce. For the island of Enggano, the only references worth mentioning are publications by Pondoppidan (1915) and De Jong (1938). Terpstra (1932) studied Sipora and the southern part of Siberut in the Mentawei Islands. Verbeek (1876) and Schröder (1917) have published reports on the island of Nias. Descriptions of rock samples of this island include those by Icke and Martin (1907).

The main source of our knowledge is contained in a number of unpublished geological reports prepared by geologists of oil companies just before the outbreak of hostilities in the Pacific: Elber (1939 a, b), Den Hartog (1940 a, b), Hopper and Dorn (1940), Koch (1941), Meyer (1941), and Todd (1939). The summaries and ample discussion of this unpublished material given by Van

Bemmelen (1949 Ia), which are accompanied by a geological sketch map, are of the utmost importance. Also of interest for our purposes are the contributions by Craandijk (1908, 1915) and Hondius van Herwerden (1910) concerning the subsidence of the western coast of the Mentawei Islands.

Topographic contour maps (1 : 100,000) based on terrestrial surveys were made of the Mentawei islands in the nineteen thirties, and wartime Japanese photogrammetric maps – also with contours and to the same scale – are available for Enggano, the Batu islands, Nias, and Simalur, as are less detailed US wartime photogrammetric maps (1 : 50,000) of Nias and the Banjak Islands. The air photos used for the compilation of these photogrammetric map series were unfortunately not at the author's disposal. His observations are based on field work carried out in 1955 on North and South Pagai, in the northern part of Siberut, and on the island of Nias.

b. Outline of the geological development

Pre-Tertiary outcrops are limited in extent. They are reported from Sipora (Mentawei Islands), a narrow zone in the Batu Islands, a small area along the southern part of the W coast of Nias, and from some localities in the Banjak Islands. The oldest rocks found on Simalur – phyllitic sandstone with quartz veins – are presumably of pre-Tertiary age.

Sediments of Tertiary age are by far the most common deposits on the islands. In broad outline, the Tertiary column consists of:

1) Lower Tertiary conglomerates and sandstones described as the 'basal sandstone series' with local occurrences of (partly pre-Tertiary?) serpentinized igneous rocks.

2) The Miocene indicated as the 'marl/limestone series'. A distinct limestone member is especially well developed in the southeastern part of Nias, and also along the W coast of Tanahmasa and on a few islets around Simalur.
In the Mentawei Islands, less resistant beds (shales, silts, and marls) predominate. The occurrence throughout the island festoon, especially in the Lower and Middle Miocene levels, of tuffaceous deposits, presumably originating from Sumatran volcanism, is interesting. The total thickness of these strongly folded sediments has been estimated at 4,000 to 5,000 m. Several unconformities can be traced, indicating relatively unimportant short interruptions of the sedimentation (temporary emergence) in a development characterized by geosynclinal subsidence.

3) A major unconformity marks the boundary between the Miocene beds and the younger, unfolded, sediments, which are composed predominantly of coralline limestones, sometimes underlain by marls. Part of the previously mentioned Miocene limestones around Simalur and on the Batu and

Banjak archipelagoes (Elber, 1939) are included in the Pliocene and Pleistocene reef cover by Den Hartogh (1940). This is, of course, a matter of considerable geomorphological interest.

The Quaternary epoch was characterized by a general uplift of the non-volcanic arc. The reef cap thus formed is especially extensive on the island of Nias, where it also reaches its maximum height of 250 to 300 m above sea-level.* River terraces and raised beach deposits indicate the latest relative uplift of 5 m, which resulted in the formation of many low plains throughout the island.

The sketch of the geological development given here is of course very general and far from complete. Our only knowledge regarding Enggano, for instance, is that strongly folded tuffaceous Neogene sediments are unconformably covered by raised coral limestones. But interesting details are known for a number of other localities, such as the intra-Neogene transverse faults with accompanying igneous rocks of Simular and the upthrusts known from the Banjak archipelago (islet of Pinang), the Batu group (between Tanahmasa and Tello), and Nias (Muzoj and Gotu upthrusts). Of geomorphological importance in the geological development sketched above, is the Quaternary uplift of the Tertiary geosyncline and the period of planation preceding it.

c. Enggano

Three geomorphological units can be distinguished on this island: the hilly interior, the raised coral reef, and the alluvial plains (including the living reefs and 5 m terraces). The latter unit (which is so small that it is mentioned only for the sake of completeness) has no noteworthy features except two long, narrow, swampy depressions – probably grabens – stretching in a north-westerly direction in the southern part of the island and penetrating into the hills (Fig. 64).

The raised coral reef surrounding the hills shows a variety of karst phenomena. It occupies almost the whole northwestern part of the island, where only one Neogene ridge emerges from it along the SW coast. The reefs reach their greatest height of about 100 m above sea-level near the highest hill of the island, Mt. Duadua (56 x 1.5; 281 m), but most of the reef cap lies below the 40 m contour line.

The hilly interior is composed of rather unresistant Neogene strata, and consequently shows intense dissection. The geological dips are predominantly northeastward, according to Pondoppidan (1913). This unit is especially well developed in the central and southwestern parts of the island. An uplifted

---

* The author gratefully acknowledges his indebtedness for information obtained from Ir. B. E. Dieperink, Chief Geologist, C.P.P.M. and from the report of Hopper and Dorn (1949).

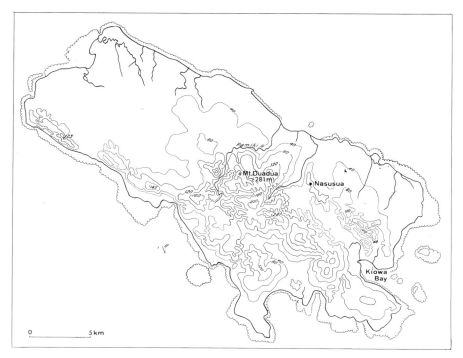

Fig. 64. Relief and drainage of the island of Enggano, reduced from the 1 : 100,000 Japanese photogrammetric map.

peneplain or summit level, such as that mentioned by Terpstra (1932) for the Mentawei Islands, has not been observed so far, although judged from the photogrammetric map its existence does not seem unlikely.

Within this unit three ridges can be distinguished, separated by the two narrow depressions already mentioned. The westernmost ridge stretches along the SW coast and reaches a maximum height of 145 m. It is narrow in the NW and broadens toward the SE. A connection with the central ridge is found near Mt Duadua, but further to the SE they are separated by one of the two narrow depressions. The central ridge is the largest and highest (Mt Duadua, 281 m). It is separated from the easternmost ridge, which is small and nowhere more than 179 m high, by the second narrow depression, probably a graben, stretching NW from the southwestern coast of the island.

A remarkable geomorphological feature is the location of the main divide in the westernmost ridge. The largest river of the island, the Pamiki, has its source there and breaks through the higher central ridge to reach the NE coast. An antecedent origin of this river course seems probable, and this implies a recent upheaval of the central ridge. It should be mentioned in this connection that the reef cap is highest in this zone. De Jong (1938) assumes recent subsidence of the coastal areas of Enggano, because of the numerous uprooted coconut trees found everywhere along the beach except on the southern part of the NE coast. This assumption accords with the subsidence observed along the eastern coasts of the Mentawei Islands (p 165). This combination of observations suggests continued warping of the island. Also interesting is the location of the island to the NE of the submarine platform, indicated on the map by the 200 m isobath, and the straight steep southwestern coast, where NE-dipping beds are cut off. It seems possible that the island once extended further to the SW and then partly subsided, but only field evidence can solve this problem.

d. The Mentawei Islands

1. *General features*

These islands are comparatively simple from a geomorphological point of view. Only two main land form units were distinguished by the author: the strongly dissected hilly landscape and the alluvium. The latter consists principally of a broad terrace lying only a few metres above the rivers, which penetrate far into the hilly unit. Mostly a sedimentation terrace, it locally shows development as an erosion terrace, especially in the upper parts of the river valleys. The small, slightly elevated coral reefs corresponding with the main river terrace as well as the few minor alluvial plains and the living reef, which are only mentioned briefly here, are of secondary importance.

The hills are generally composed of rather unresistant rocks, and there is consequently an extremely high rate of normal erosion. Most of the ridges and divides have thus been reduced to very steep crests, and few flat-topped hills occur. The hills and ridges rise almost without transition from the broad river terrace. The strike of these weak beds apparently had little significance for the river system; the direction of the ridges is almost completely independent of the geological structure, although of coourse there are some exceptions. This independence was observed by Terpstra (1932), and is confirmed by the present author's observations on North and South Pagai and the northern part of Siberut. The impression is gained that a planation surface was formed on the islands, after the actual folding of the Tertiary beds and prior to the uplift, that gave rise to the present island festoon. This hypothesis is supported by the characteristics of the relief mentioned above, and particularly by the local similarity in the height of hill tops and ridges, which points to the existence of a summit level. The higher parts of the hilly areas are distinguished on the geomorphological map from the lower areas, but this is not meant to suggest two summit levels at different heights, for which no proof was found; it seems rather to be an indication of different degrees of uplifting of parts of one and the same level. The few hills rising above this severely eroded and dissected level are best explained as monadnocks, since they consist of amphibolites, serpentines, quartzitic sandstones, etc.

The most remarkable geomorphological phenomenon shown by the islands is the absence of higher elevated reefs, such as those known from Enggano, Nias, etc. This may point to a different tectonic development of the Mentawei group, as compared to the other parts of the island arc, during the Quaternary uplift. The peneplain and the broad low river terrace occurring throughout the island festoon offer an indication that at least the early and most recent stages of this uplift were more or less the same. A mere difference in time of the uplift would in itself be sufficient to explain the lack of reefs, since there may be no period of relative standstill if sea-level changes and diastrophism alternate. The lack of evidence of the existence of higher elevated reefs in the Mentawei group might also be attributed to unfavourable circumstances for coral growth, possibly caused by a high silt content of the sea water, resulting from rapid erosion of the weak rocks of the islands. This situation would certainly offer an interesting subject for future geomorphological field investigations, which should be based on interpretation of aerial photographs.

It is interesting to note that the physical setting is distinctly reflected in the pattern of human activities on the islands. Agriculture is confined to the river terraces, where the young soils are evidently rather fertile and yield good harvests; no gardens occur on the steep hills. The rapid erosion and the very high silt content of the rivers are definitely due not to man but to the weakness of the sediments concerned. Wet rice cultivation has recently started in the mostly swampy shallow moats of the low reefs mentioned above. The lithology of the islands also accounts for the rare occurrence of rapids, making the rivers easily navigable for the native canoes. Because there is a scarcity of material for vad construction, transportation is almost exclusively by canoe on the rivers except on the more densely populated island of Nias, where raised reefs and limestones render the rivers mostly unnavigable and a road system has been developed. On the other islands there are only a few trasil crossing the devides and along the coasts. The rather straight western coasts are exposed to the strong surf of the Indian Ocean, and are therefore very sparsely populated.

2. *South and North Pagai and Sipora*

The island of South Pagai is formed almost completely of hills having a rather uniform height (200–300 m), i.e. the dissected planation surface mentioned earlier. Only the narrow peninsula forming the southernmost part is much lower. The alluvium is best developed along the eastern side of the hilly area, and along the SE coast (minor occurrences are also found along the central part of the W coast and near the northern tip of the island). Fig. 65, a sketch of the NE coast near Simakalo, shows the two land form units, *viz.* the dissected hills with their summit level and the bordering allu-

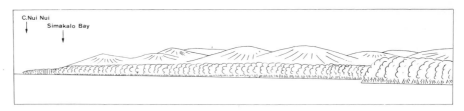

Fig. 65. Field sketch of the summit level, indicating an uplifted and dissected 'peneplain', of South Pagai (Mentawei Islands). The mangroves in the foreground indicate a slightly uplifted coastal terrace or coral reef.

vium along the coast, which is often covered with mangroves. This picture is typical of the whole Mentawei group.

The main divide of South Pagai generally lies quite close to the W coast. The southern part of the island is characterized by a somewhat trellis-type drainage pattern certainly influenced by the geological structure, a rather uncommon feature in the archipelago.

North Pagai is separated from the southernmost island by the narrow Sikakap Straits, which may well have originated from recent faulting. Serpentinized rocks crop out at various places along its coasts. Slightly elevated coral reefs also occur at about 2–3 m above sea-level. The highest parts of the hilly interior of the island form a zone which is about 200–300 m high and crosses the island more or less diagonally from the southeastern end to the middle of the W coast. In the northern tip the summit level is much lower (max. 100 m) on both sides. The southwestern part of the island is separated from the higher zone by the broad valley of the Sikako River, where terraces are beautifully developed at a maximal height of 5 m above the river. The main divide is located at a short distance from the W coast.

The lower parts of the summit level of Sipora are rather limited in area, and are located in the extreme S and along the N and NE coasts. The broad valley of the straight Saurenu River is an important element. Except at the southern end of the island the divide is in the SW, although the highest hills nevertheless show a distinct preference for the northeastern side of the island, where the amphibolites occur.

3. *Siberut*

The northernmost island, Siberut, is the largest and highest (384 m) of the Mentawei archipelago. A trellis drainage pattern showing the influence of geological structure on the river development is more distinct here than elsewhere. The main divide is not generally located on the southwestern side of the island. Swampy lowlands occur frequently along the indented N and NE coasts, but are unimportant along the straight SW coast. Broad river terraces, sometimes hundreds of metres wide, are observed at a level of 5 to 6 m everywhere in the northern part of Siberut. These are mainly sedimentation terraces in which distinct cross-bedding occurs locally. In several localities along the upper course of the rivers, however, the Tertiary beds are cut off by a terrace, thus indicating the formation of the latter by river erosion. Locally, there are lower terraces lying 1.5 to 2 m above the river.

The W coast of the island was studied at the mouth of the Simalegi River and somewhat further S, near Cape Gobdjib. A narrow alluvial plain with a beach ridge along the shore is present almost everywhere and reaches a height of about 1 to 2 m above sea-level. This same plain could be traced in the surroundings of Lake Gobdjib. We assume that this level corresponds with the 5 m terrace along the rivers, because in two river valleys this terrace decreases rather suddenly to a height of about 2 m, which indicates a certain degree of warping of the coastal area of the island. This general warping probably occurred more generally in the archipelago and would easily account for the fact that the slightly elevated reefs observed, for example, on North and South Pagai, are always less than 5 m high.

4. *Submergent coasts and islands to the East*

Subsidence is more pronounced along the E coasts of the islands and especially Siberut, as already suggested by the greater size of the alluvial plains there. Numerous isolated hills of Tertiary rocks emerge from the swampy plain, and subsidence of these coastal areas is evident (Fig. 66). Although one feels inclined to attribute the land forms to the post-Glacial rise in sea-level, the absence of the 5 m terrace in these areas is proof that tectonic subsidence still continues and doubtless also influences the development of

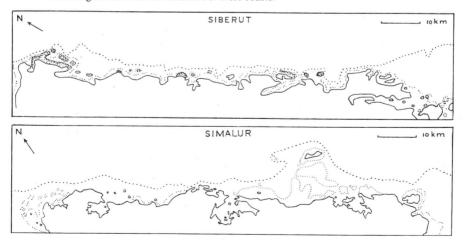

Fig. 66. The eastern coasts of the islands of Siberut and Simalur, as they appear on a chart scaled 1 : 100,000. Note the irregular shoreline with many bays, capes, and islets, reflecting the dominant subsidence of these coasts.

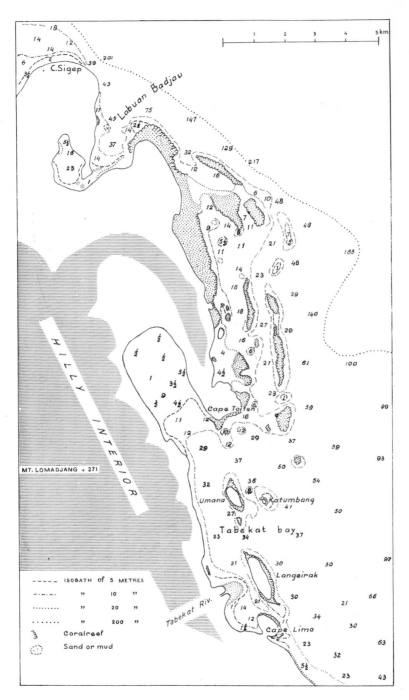

Fig. 67. Tabekat Bay and nearby reefs and islets in the northeastern part of Siberut, as depicted on a chart. Note the Cape Limo tombolo and the pseudo-barrier reef structure. Scale 1 : 125,000.

the coastal areas. Recent tectonics are also evident from the frequent earthquakes with accompanying sea waves (*tsunami*). Hondius van Herwerden (1910) and Craandijk (1915) mention transgression, resulting in complete submergence of a number of small islets off the E coast of the Mentawei group, as evidence of recent diastrophism in this area, which was confirmed by statements made by the local population.

Other coastal phenomena point in the same direction. Fig. 67 illustrates the coastal area of NE Siberut around Tabekat Bay. The seaward side of this bay is formed by a row of coral islands and reefs running from the tombolo of Cape Limo via the islands of Langeirak and Umana/Katumbang to Cape Toiten in the N. The subsoil of these islands certainly represents a ridge of Tertiary sedimentary rocks emerging in the Toiten peninsula. The bay is actually the lagoon of a pseudo-barrier reef reaching a maximum depth of 32 m. At the seaward side of the Toiten peninsula two similar reef structures can be seen, their ridges separated by lagoons with a depth of 18 and 27 m. The field sketch in Fig. 68 shows the bay as seen looking northward from the water W of Cape Limo.

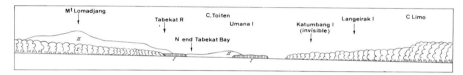

Fig. 68. Field sketch of Tabekat Bay on the E coast of Siberut, looking N. The row of islands (Umana, Katumbang, and Langeirak) indicates the edge of the pseudo-barrier reef, the lagoon of which is formed by Tabekat Bay. (*cf.* Fig. 67).

The mouth of the Sikabaluan River is located S of Cape Limo, and the spread of the delta of this river in the lagoon obscures the subsidence there. The coral reef present at the seaward tip of the delta is the only indication of the presence of the pseudo-barrier reef. Further to the S, recent abrasion is again evident in the continuation of the lagoon. The whole E coast of the island is apparently subsiding, and the same applies to the other islands, where pseudo-barrier reefs also exist, for instance off the NE coast of South Pagai.

It should be noted in this connection that off the W coast of central and northern Sumatra there is a pseudo-barrier reef (see geomorphological map) which can be traced S of the Batu Islands and is almost completely connected with the pseudo-barrier reef of NE Siberut. Although the depth of its lagoon is somewhat greater locally, there are no essential differences. Molengraaff (1922) attributed the West-Sumatran barrier reef mainly to the post-Glacial rise in sea-level and the high rate of upward coral growth observed also elsewhere in Indonesia. However, the lack of a 5 m reef terrace related to the sub-Recent lowering of the sea-level shows that in more recent diastrophism has also played an essential role. We may therefore safely assume that the part of the interdeep formed by the Mentawei basin is still actively subsiding.

e. The northern part of the Island Festoon

1. *Batu Archipelago*

Two of the three larger islands of the Batu group, *viz.* Tanahbala and Tanahmasa, form the continuation of the non-volcanic arc and link the Mentawei Islands (Siberut) with the island of Nias. The third island, Pini, is located further to the E on the transverse ridge of the interdeep connecting Sumatra with the non-volcanic arc.

Geomorphologically the island of Pini is rather similar to the lower part of the 'interior hills' of the Mentawei group. The greatest height, 80 m, is reached in the eastern part, which is separated from the western part (60 m) by a lower central belt. Alluvium is concentrated mainly along the N and E coasts. The picture presented by the rivers, coasts, and small surrounding islets, suggests the presence of two rectilinear features, possibly faults.

By far the greater part of the island of Tanahbala lies within the same geomorphological unit. The highest area reaches 270 m. A narrow zone in the E, bordering Tanahbala Straits, is divergent, however, being characterized by a general westward slope, separated from the rest of the island by a belt of somewhat lower relief, and nowhere reaching a height of more than about 100 m. Further to the N, this zone is submerged except for a few small islands.

On the other side of the Tanahbala Straits, the long, narrow island of Tanahmasa (max. height 204 m) shows the same characteristics as the adjacent part of Tanahbala, but there the general slope of the terrain is directed eastward. The impression is gained that the Tanahbala Straits represent a longitudinal graben formed in the axis of the up-arched zone. Tanahmasa is indicated as an anticlinal feature on the geological map (Van Bemmelen, 1949); according to the present author, this might be interpreted as a huge flexure. Such features are also known from other localities in island-arc areas. Uplift of the zone is indicated by the raised coral reefs reported from the easternmost part of Tanahbala.

The W coast of Tanahmasa is formed by a limestone ridge of the marl-limestone series mentioned in the discussion of the geology of the non-volcanic arc. This resistant bed can be traced further to the N, where it strongly influences the land forms of Nias. According to Kissling (1948), the marl-limestone series occur particularly in the northern part of the island festoon, where they form the flanks of the geanticline. The central parts of the arc are mainly composed of the tuff-marl series.

2. Nias

This is the largest and highest island of the non-volcanic arc; it is also the least simple geomorphologically. The hilly interior can be subdivided as follows.

The greatest height is reached in the southern half, where the general elevation is about 500 to 600 m and the highest hill (Lolomatur, 886 m) is located. The northern part of the island is distinctly lower, reaching 200 to 300 m. Most of the terrain is intensely dissected, and the drainage pattern is strongly influenced by the two upthrusts already mentioned (p 160). The Eho River runs along the Gotu upthrust in the southwestern part of the island, and the Ojo and Muzoj rivers are located at the foot of the Muzoj upthrust, which is geomorphologically even more important. Both upthrusts lie more or less parallel to the longitudinal axis of the island.

A further sub-division is based on the well-marked difference in relief between the areas where low resistance marls, etc., crop out and the part where harder rocks, especially sandstones, are found. The former area is always low and greatly resembles the 'hilly interior' of the Mentawei Islands, whereas the latter area has a higher relief and is more coarsely dissected, a summit level being almost indistinguishable. The two fragments of the Japanese photogrammetric map of the island shown in Fig. 69 illustrate this.

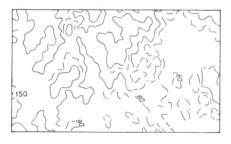 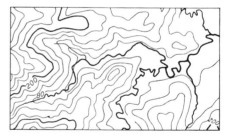

Fig. 69. Two fragments of the Japanese photogrammetric map of Nias to scale 1 : 100,000.
Left: gently rolling hills formed in marls. Right: area of higher relief formed in sandstones.

The most striking geomorphological feature of Nias is the conspicuous ridge of Miocene limestone running parallel to the E coast. Marls and other rocks of low resistance crop out to the W of it, thus forming a depression at its landward side. The ridge stands out clearly in relief and the rivers break through it in steep narrow gorges, the best example being given by the gorge shown in Fig. 70. Conical karst hills are a common feature of this ridge, for instance near the town of Gunungsitoli (Fig. 69). Verbeek (1876) described these rocks as young coral limestone, but the palaeontological investigations

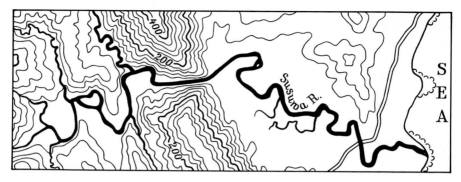

Fig. 70. Fragment of the Japanese photogrammetric map of Nias to scale 1 : 100,000, showing the narrow short gorge of the Susuwa River where it breaks through the limestone ridge in the eastern part of the island.

of Icke and Martin (1907) proved that at least part of the limestones is of Miocene age, an opinion supported by more recent investigators. The view of Schröder (1907), who interpreted the phenomenon as a raised barrier reef, must therefore be rejected.

Nias has rather large raised coral reefs, especially in the northwestern tip and along the Muzoj lowlands, which clearly must once have formed a long bay penetrating into the northern part of the island at the foot of the Muzoj upthrust. Other occurrences are formed SE of the Ojo River and along the southern end of the E coast. They cover about 15 per cent of the island area and ara characterized by riverless plateaus with conical karst hills. Downstream from these areas the rivers often show active travertine formations. This partial reef cap, and also the brackish water fauna of the Hilina cave, point to an uplift of at least 130 m. No certain indications of such higher reefs have been reported along the E coast near the above-mentioned limestone ridge, but scattered remnants possibly occur there. Hopper (1940)

Fig. 71. Field sketch of the surroundings of Gunungsitoli, showing the 5 m coral terrace and conical karst hills formed in the limestone ridge of eastern Nias.

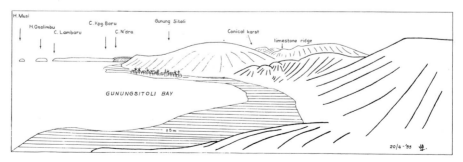

mentions gentle seaward dips of the reef, resulting from deformation. This would indicate a recent further warping of the island, which is consistent with the frequent occurrence of earthquakes. The original slope of the reef should be taken into account, however; further studies will be necessary before warping can be demonstrated.

The third geomorphological unit, the alluvium, is well developed along the Muzoj River, along parts of the W coast, and especially in the remarkable and large outward bulge on the E coast in the southern part of the island. The 5-m reef and river terraces, which also appear in this unit, are beautifully developed in many localities throughout the island, for instance near the town of Gunungsitoli (Fig. 71). The river terraces are of course generally less broad than the extremely large ones described from the Mentawei Islands, because more resistant rocks predominate on Nias. The 5 m terrace is attributed to a lowering of the sea-level. This suggests that the uplift of the coastal areas came to a standstill in recent times.

3. *Banjak Islands and Simalur*

The Banjak Islands are located on the northernmost transverse threshold across the interdeep connecting Sumatra and the non-volcanic arc. Their relief corresponds in general to the 'hilly interior' unit of the other islands. The distinct limestone ridge (mentioned e.g. for Nias) stands out in clear relief along the N coast of the island of Tuangku as well as in a small area of Udjungbatu, and includes the highest elevation reached in the group, 313 m. Raised coral reefs also occur, for example in a large part of Udjungbatu.

The northernmost island of the arc, Simalur, has a rather fragmented character due to the many transverse faults, some of which are indicated on the map. The numerous bays are certainly also due to these faults (Fig. 66). The 'hilly interior' occupies most of the island. It is relatively low as compared, for example, with Nias, except in the southern part, where a higher zone (max. 567 m) runs across it from E to W. The limestone ridge just mentioned is reported from a few small surrounding islands only. The extreme southeastern tip is formed by a raised coral reef, probably reaching a height of 100 to 120 m. A number of long narrow alluvial plains forming the continuation of bays and lying parallel to the geological structure, are noteworthy in the northern part of the island. Other occurrences of alluvium are found along the NW coast.

Recent subsidence of the E coast was suggested by Craandijk (1908), who believed that the islet of Simalur Tjut had disappeared completely in 1906. The recent development of this coast is apparently rather similar to that of the sinking E coast of the Mentawei Islands (Fig. 66). The (pseudo) barrier reef off the W coast of Sumatra can be traced as far N as Nias and Simalur. The lack of a 5 m reef terrace in this barrier reef zone points to recent subsidence of this part of the interdeep, which has also been mentioned (p 170) for the central-Sumatran section of the interdeep.

IX Bibliography

ADAM, J. W. H. – 1932/'33 – Kaksa Genese. De Mijningenieur, 13, pp. 217–221; 14, pp. 1–5; 20–26; 81–87; 167–171.

ALEVA, G. J. J. – 1960 – The plutonic igneous rocks from Billiton. Geologie en Mijnbouw, 39, pp. 427–436.

BAARTMANS, J. A. et al. – 1947 – De morfologie van de Java- en Soenda Zee. Tijdschr. Kon, Ned. Aardr. Gen., 64, pp. 442–465; 555–576.

BADINGS, H. H. – 1937 – Het Palaeogeen in den Indischen Archipel. Verh. Kon. Ned. Geol. Mijnb. Gen., Geol. Serie, 12, pp. 233–299.

BALAZS, D. – 1968 – Karst Regions in Indonesia. Karstz-és-barlangkutatás, V., Budapest.

BEHRMAN, W. – 1921 – Oberflächenformen in den feutchtwarmen Tropen. Ztschr. d. Ges. f. Erdkunde, Berlin, 30, pp. 44–60.

BELTZ, E. W. – 1944 – Principal sedimentary basins in the East Indies. Bull. Am. Ass. Petrol. Geol., 28, pp. 1140–1454

BEMMELEN, R. W. van – 1929 a – The origin of Lake Toba (N. Sumatra). 4th Pac. Sci Congr. Bandung, Proc. IIA, pp. 115–125.

BEMMELEN, R. W. van – 1929 b – Nota betreffende een geologisch onderzoek van den Asahan weg (Porsea-Bandarpulau). Unpubl. report Dienst v.d. Mijnbouw N.I.

BEMMELEN, R. W. van – 1930 – Geologische waarnemingen in de Gajo-landen (N. Sumatra) Jaarb. Mijnw. N.I., Verh. III, pp. 71–94.

BEMMELEN, R. W. van – 1931 – Het Boekit Mapas-Pematang Semoet vulkanisme (Z. Sumatra). Verh. Kon. Ned. Geol. Mijnb. Gen., Geol. Serie, 9, pp. 57–76.

BEMMELEN, R. W. van – 1932 a – Een voorbeeld van winst en verlies van stroomgebied door een vulkanische eruptie. De Mijningenieur in N.I., 13, pp. 212–213.

BEMMELEN, R. W. van – 1932 b – Geologische kaart van Sumatra schaal 1 : 200,000. Toelichting bij blad 10 (Batoeradja). Dienst v. d. Mijnbouw N.I.

BEMMELEN, R. W. van – 1933 – Geologische kaart van Sumatra schaal 1 : 200,000. Toelichting bij blad 6 (Kroeï). Dienst v. d. Mijnbouw N.I.

BEMMELEN, R. W. van – 1934 – De tektonische structuur van Z. Sumatra. Natuurk. Tijdschr. N.I., 94, pp. 7–14.

BEMMELEN, R. W. van – 1935 – Vulkano-tektonische depressies op Sumatra. Hand. 25ste Ned. Nat. en Geneesk. Congres, Leiden, pp. 289–296.

BEMMELEN, R. W. van – 1939 – The volcanic-tectonic origin of Lake Toba (N. Sumatra). De Ingenieur in N.I., 6, pp. 126–140.

BEMMELEN, R. W. van – 1949 – The Geology of Indonesia (2Vols.). The Hague, Govt. Print, 732 + 265 pp.

BEMMELEN, R. W. van and P. ESENWEIN – 1932 – De liparietische eruptie van den bazaltischen Tanggamoes vulkaan. Wetensch. Meded. Dienst v.d. Mijnbouw N.I., 22, pp. 33–62.

BEMMELEN, R. W. van and J. ZWIERZYCKI – 1936 – Het Palaeogeen op Sumatra. De Ingenieur in N.I., 4, Section IV, pp. 9–29.

BEMMELEN, W. van – 1919 – De Piek van Kerintji. Natuurk. Tijdschr. N.I., 78, pp. 173–184.

BRAAK, C. – 1925/'29 – Het Klimaat van Nederlandsch Indië. Kon. Magn. en Meteor. Obs. Batavia, Verh. 8.

BROUWER, H. A. – 1914 – Bijdrage tot de geologie van de Boven-Kampar en Rokanstreken. Jaarb. Mijnwezen N.I., 43, Verh. 1, pp. 130–170.

BROUWER, H. A. – 1915 – Erosie verschijnselen in de tuffen van de Padangsche bovenlanden. Tijdschr. Kon. Ned. Aardr. Gen., XXXII, pp. 338–345.

BROUWER, H. A. – 1925 – The Geology of the Netherlands Indies.

BRUYN, J. W. de – 1951 – Isogam maps of the Carribbean Sea and surroundings and of Southeast Asia; Proc. 3rd World Petroleum Congress, pp. 598–612.

BÜCKING, H. – 1904 – Zur Geologie von Nord- und Ost-Sumatra. Samml. Reichsm. Geol. Museum, Leiden, 8.

BURTON, C. K. – 1962 – The older alluvium of Johore and Signapore. Fed. of Malaya Geol. Survey, prof. paper E-62.2 G.

COLLETTE, B. J. – 1954 – On the gravity field of the Sunda region (W. Indonesia). Geol. en Mijnbouw, new series, 16, pp. 271–300.

COLLINGS, H. D. – 1938 – Pleistocene site in the Malay Peninsula. Nature, 142, pp. 575–576.

COURTIER, D. B. – 1962 – Note on terraces and other alluvial features in parts of Province Wellesley, S. Kedah and N. Kedah. Fed. of Malaya Geol. Survey, prof. paper E-62. 1T.

CRAANDIJK, C. – 1908 – Onnauwkeurigheden in de pas verschenen kaart der Topografische Inrichting te Batavia. Tijdschr. Kon. Ned. Aardr. Gen. XXV, pp. 581–583.

CRAANDIJK, C. – 1915 – Verandering in eilandjes en riffen op de Oostkust der Mentawei-eilanden (Westkust Sumatra). Tijdschr. Kon. Ned. Aardr. Gen. XXXII, pp. 652–653.

DOOP, J. E. A. de – 1922 – Deli en de Karo-hoogvlakte als Lahar-product. Natuurk. Tijdschr. N.I., 82, pp. 208–214.

DERSJANT, V – 1926 – Het kalksintercomplex Dolok Tinggi Radja (S.O.K.). Jaarversl. Top. D., p. 125.

DRUIF, J. H. – 1932/'39 – De bodem van Deli (I–IV). Meded. Deli Proefstation.

DRUIF, J. H. – 1935 – Over gesteenten van Poeloe Berhala (S.O.K.). Proc. Kon. Ned. Akad. v. Wetensch. A'dam, 38, pp. 639–649.

DRUIF, J. H. – 1943 – Verslag van een verkenning van het gebied rondom het Goemai gebergte, Res. Palembang. Unpubl. Report Bodemk. Inst. Bogor.

DURHAM, J. W. – 1940 – Oeloe Aer fault zone, Sumatra, Bull. Am. Ass. Petrol. Geol., 24, pp. 359–362.

EARLE, W. – 1845 – On the physical structure and arrangement of the islands of the Indian Archipelago. J. Royal Geogr. Soc. London, 15, pp. 358–365.

EDELMAN, C. H. – 1941 – Studiën over de bodemkunde van Nederlandsch-Indië. Publ. Fonds Landb. Export Bureau 1916–1918, NR. 24.

ELBER, R. – 1939 a – Bericht über eine regional-geologische Untersuchung von Nias. Unpubl. Report B.P.M.

ELBER, R. – 1939 b – Geologische Übersicht über die Inseln Simalur, Banjak Gruppe, Nias, Batu Gruppe. Unpubl. Report B.P.M.

ERB, J. – 1905 – Beiträge zur Geologie und Morphologie der südlichen Westküste von Sumatra. Ztschr. d. Ges. f. Erdkunde, Berlin, 14, pp. 251–281.

ES, L. J. C. van – 1918 – Geologische overzichtskaart van den Ned. Oost Indischen Archipel 1 : 1.000.000, Toelichting bij blad XV.

FABER, D. A. – 1954 a – Verslag van een bodemkundige verkenning van het eiland Bangka. Unpubl. Report Balai Penjelidikan Tanah, Bogor.

FABER, D. A. – 1954 b – Verslag van een bodemkundige verkenning van het eiland Billiton. Unpubl. Report Balai Penjelidikan Tanah, Bogor.

FENNEMA, R. – 1887 – Topografische en geologische beschrijving van het noordelijk gedeelte van het Gouvernment S.W.K. Jaarb. Mijnwezen N.I., Verh. I., pp. 129–252.

FRIJLING, H. - 1925 - Geologisch-Mijnbouwkundig onderzoek in den omtrek van de Asahanen Koeala rivieren (Tobalanden, O. Sumatra) Jaarb. Mijnwezen N.I., Verh. II.

GERMERAAD, J. H. - 1941 - On the rocks of the isle of Koendoer, Riouw Archipelago, N.E.I. Proc. Kon. Ned. Akad. v. Wetensch., A'dam, 44, pp. 1227-1233.

HAAN, W. de - 1935 - Gesteenten van Sumatra's Westkust. De Ingenieur in N.I., 2, IV, pp. 88-97.

HAGEN, F. - 1924 - Een bestijging van den G. Berapi. Jaarverslag. Top. D. N.I., XX, pp. 59-60.

HARTMANN, M. A. - 1936 - De Hoeloebeloe hoogvlakte in Z. Sumatra. De Tropische Natuur, XXV, pp. 53-60.

HARTOG, L. E. W. den - 1940 a - Geological reconnaissance of Pulau Siberut. Unpubl. Report N.P.P.M.

HARTOG, L. E. W. den - 1940 b - Geological reconnaissance of Pulau Simalur. Unpubl. Report N.P.P.M.

HEIM, A. - 1929 - Beobachtungen auf dem Merapi Sumatras in August 1928. Ztschr. f. Vulkank., 12, pp. 322-326.

HERWERDEN, J. H. Hondius van - 1910 - Het dalen van eilandjes op de Oostkust der Mentawei-eilanden (bewesten Sumatra). Tijdschr. Kon. Ned. Aardr. Gen., XXVII, pp. 792-793.

HETZEL, W. H. - 1939 - Waduk Djapara. Unpubl. Report Geol. Mijnb. Dienst, Bandung.

HEURN, F. C. van - 1923 - Studiën betreffende den bodem van Sumatra's Oostkust, zijn uiterlijk en zijn ontstaan.

HIRSCHI, H. - 1910 - Geographisch-Geologische Skizze des Nordrandes von Sumatra. Tijdschr. Kon. Ned. Aardr. Gen., CCVII, pp. 741-763.

HOEN, C. W. A. 't - 1929 - De Tertiaire petroleum terreinen ter Oostkust van Sumatra (Atjeh II) Jaarb. Mijnwezen N.I., Verh. I.

HOOZE, J. H. J. - 1878 - Verslag over de artesische drinkwater-voorziening in Groot-Atjeh. Jaarb. Mijnwezen N.I., pp. 3-52.

HOPPER, R. H. and C. L. DORN - 1940 - A geological reconnaissance in Western and Northern Nias. Unpubl. Report N.P.P.M.

HORSTING, L. H. C. - 1927 - De vulkaan Sibajak. Ned. Indië oud en nieuw, XII, pp. 85-94.

ICKE, H. and K. MARTIN - 1907 - Over tertiaire en kwartaire vormingen op het eiland Nias. Samml. Reichsm. Geol. Leiden, 8, pp. 204-252.

IDENBURG, A. G. A. - 1937 - Systematische grondkarteering van Zuid Sumatra. Thesis Wageningen.

JONGH, A. C. de - 1929/'30 - Inleiding tot eene systematiek der Sumatra gronden. Alg. Landb. Weekbl. N.I., 14, II, pp. 955-958; 987-990.

JAARVERSLAG. v. d. TOP. DIENST in N.I. - 1904/'40 - Various cratermaps. (Boer ni Tèlong: 1910, pp. 86-88; Dempo: 1907, pp. 59-61; 1910, pp. 78-79; Gedang-Daun: 1915, pp. 121-126; Kaba: 1907, pp. 61-64; 1914, pp. 149-169; Kerintji: 1924, pp. 59-60; Kunjit: 1925, p. 95; Pasagi: 1911, pp. 101-104; Pasawaran-Ratai: 1903, p. 69; Radjabassa: 1908, p. 79; Sekintjau-Belirang: 1911, pp. 74-78; Sibajak: 1914, pp. 170-173; Singgalang/Tandikat: 1928, pp. 121-122; Sumbing/Labuh: 1925, pp. 98-99; Tauggamoes: 1907, pp. 68-69; 1912, pp. 107-108; Toedjoeh: 1925, pp. 133-134.

JOCHEM, S. C. J. - 1929 - De kalksinterterrassen bij den Tinggi Radja op de Oostkust van Sumatra. De Tropische Natuur, 10, pp. 21-30.

JONG, J. K. de - 1938 - Een en ander over Enggano. Natuurk. Tijdschr. N. I., 98, pp. 3-46.

JUNGHUHN, F. - 1853/'54 - Java, deszelfs gedaante, bekleeding en inwendige structuur.

KARACSON, A. A. von - 1897 - De vulkaan Kaba op Sumatra. Tijdschr. Kon. Ned. Aardr. Gen., XIV, pp. 555-570.

KATILLI, J. A. - 1968 - Cross-folding in Bangka, West Indonesia. Contrib. Dept. Geol. ITB, 68, pp. 61-70.

KATILLI, J. A. – 1969 – New results of radiometric age dating of some Indonesian Quaternary deposits. Bull. NIGM Bandung, 2, pp. 29–31.

KATILLI, J. A. – 1970 a – Large transcurrent faults in SE Asia with special reference to Indonesia. Geol. Rundschau, 59, pp. 581–600.

KATILLI, J. A. – 1970 b – Naplet structures and transcurrent faults in Sumatra. Bull. NIGM Bandung, 3, pp. 11–28.

KATILI, J. A. and F. HEHUWAT – 1967 – On the occurrence of Large Transcurrent Faults in Sumatra, Indonesia, J. of Geosciences Osaka, X, pp. 5–17.

KATILLI, J. A. and H. D. TJIA – 1968 – Outline of Quaternary tectonics of Indonesia. Bull. NIGM Bandung, 2, pp. 1–10.

KEMMERLING, G. G. L. – 1920 – Vulkanen en vulkanische verschijnselen in de Residentiën Sumatra's Westkust (N. deel) en Tapanoeli. Vulk. Meded. 1, Dienst Mijnwezen N.I.

KINT, A. – 1935 – De luchtfoto en de topografische terreingesteldheid in de mangrove. Visser & Co, Djakarta, 18 pp.

KISSLING, E. A. – 1948 – Enkele stratigrafische mededelingen over de eilanden voor de ZW-kust van Sumatra. Geol. en Mijnbouw, 10, p. 118.

KLEIN, W. C. – 1918 – De Oostoever van het Toba-meer in Noord-Sumatra, Jaarb. Mijnwezen N.I., 46, Verh. I, pp. 136–187.

KLOMPE, Th. H. F. – 1955 – On the supposed upper Palaeozoic unconformity in N. Sumatra. Leidse Geol. Meded., 20, pp. 120–134.

KLOMPE, Th. H. F. – 1957 – The status of geological mapping in Indonesia. Contr. Dept. of Geol., Univ. of Indonesia, 31.

KROL, G. L. – 1960 – Theories on the genesis of 'kaksa'. Geol. en Mijnbouw, 39, pp. 437–443.

KUGLER, H. – 1922 – Geologie des Sangir-Batanghari Gebietes (Mittel-Sumatra). Verh. Geol. Mijnb. Gen., Geol. Ser. 5, pp. 135–201.

LEHMANN, H. – 1933 – Kultur-geographische Wandlungen in Südost-Sumatra. Zeitschr. d. Ges. f. Erdk. Berlin, 42, pp. 161–175.

LINGER, H. – 1941 – Der Vulkan Dempo in Sud Sumatra. Die Alpen, XVII, pp. 231–237.

LOHUIZEN, H. J. van – 1924 – Geologisch onderzoek van een gedeelte van het landschap Langkat (S.O.K.). zjaarb. Mijnwezen N.I., Verh. I.

MAREL, H. W. van der – 1947 – Diatomeënaarde afzettingen in de omgeving van het Tobameer. De Ingenieur, 59, Afd. Mijnb., pp. 58–63.

MAREL, H. W. van der – 1948 a – Het Tobameer. Geol. en Mijnbouw, 10, pp. 80–89.

MAREL, H. W. van der – 1948 b – Volcanic glass, allanite and zircon as characteristic minerals of the Toba Rhyolite Tuff at Sumatra's East Coast. J. of Sed. Petr. 10, pp. 24–29.

MEYER, H. – 1941 – Nota over een aanvullend geologisch onderzoek op het eiland Nias. Unpubl. Report B.P.M.

MOERMAN, C. – 1915 – Verslag van een geol. mijnbouwk. verkenningstocht in een gedeelte der Residenties Benkoelen en Palembang. Jaarb. Mijnwezen N.I., Verh. I.

MOHR, E. C. J. – 1922 – De grond van Java en Sumatra.

MOHR, E. C. J. – 1933/'38 – De bodem der tropen in het algemeen en die van Nederlandsch Indië in het bijzonder. Meded. XXXI, Kol. Inst., Afd. Handelsmus., 12 (6 vols).

MOLENGRAAFF, G. A. F. – 1929 – The coral reefs in the East Indian Archipelago, their distribution and mode of development. Proc. IVth Pac. Sci. Congr., Vol. 2, pp. 55–89.

MOLENGRAAFF, G. A. F. and M. WEBER – 1919 – Het verband tusschen den plistocenen ijstijd en het ontstaan der Soendazee en de invloed daarvan op de verspreiding der Koraalriffen en op de land- en zoetwater fauna. Versl. Kon. Akad. v. Wetensch. A'dam, Afd. Wis. en Nat., 28, pp. 497–544.

MONTAGNE, D. G – 1950 – Het terrassenlandschap van de Atjehvallei tusschen Seulimeum en Indrapuri, gezien vanuit de lucht. Instituut Geografi Djatop., Publ. 3.

MÜLLER, J. – 1895 – Nota betreffende de verplaatsing van eenige triangulatiepilaren in de Residentie Tapanoeli t.g.v. de aardbeving van 17 mei 1892. Nat. Tijdschr. N.I., 54, pp. 299–309.

MUSPER, J. – 1927 – Indragiri en Pelalawan, uitkomsten van een geol. mijnbouwk. onderz. Jaarb. Mijnwezen N.I., Verh. I, pp. 1–247.

MUSPER, J. – 1930 – Beknopt verslag over uitkomsten van nieuwe geologische onderzoekingen in de Padangsche Bovenlanden. Jaarb. Mijnwezen N.I., Verh., pp. 265–331.

MUSPER, J. – 1934 – Een bezoek aan de grot Soeroeman Besar in het Goemaigebergte, Z. Sumatra. Tijdschr. Kon. Ned. Aardr. Gen., 51, pp. 521–531.

MUSPER, K. A. F. R. – 1933 a – Geologische kaart van Sumatra, schaal 1 : 200,000. Toelichting bij blad 15 (Prabamoelih). Dienst v. d. Mijnbouw N.I.

MUSPER, K. A. F. R. – 1933 b – Geologische kaart van Sumatra, schaal 1 : 200,000. Toelichting bij blad 16 (Lahat) Dienst v. d. Mijnbouw N.I.

NEUMANN VAN PADANG, M. – 1940 – Shifting Craters of the Talakmau Volcano, Sumatra, J. of Geom., 3, pp. 218–226.

NEUMANN VAN PADANG, M. – 1951 – Catalogue of the active volcanoes of the world, including solfatara fields. Part I, Indonesia.

NEVE, G. A. de – 1956 – Combined air- and terrain-reconnaissances carried out by the Volc. Survey in Indonesia. Proc. 8th Pac. Sci. Congr., Vol. II, pp. 177–178.

OBDEYN, V. – 1941/'43 – Zuid-Sumatra volgens de oudste berichten I–III. Tijdschr. Kon. Ned. Aardr. Gen., 58, pp. 190–217; pp. 322–342; 476–503; 59, pp. 46–76; pp. 742–771; 60, pp. 102–111.

OPPENOORTH, W. F. F. and J. ZWIERZYCKI – 1918 – Geomorphologische en tektonische waarnemingen als bijdrage tot de verklaring van de landschapsvormen van N. Sumatra. Jaarb. Mijnwezen N.I., Verh. I, pp. 276–311.

OVEREEM, A. J. A. van – 1960 – The geology of the cassiterite placers of Billiton (Indonesia), Geol. en Mijnbouw, 39, pp. 444–457.

PANNEKOEK, A. J. – 1938 – Terreinvormen in de tropen. Hand. N.I. Natuurw. Congr. Soerabaja, pp. 467–469.

PHILIPPI, H. – 1916 – Morphologische en geologische aanteekeningen by de kaart van Z. Sumatra. Jaarversl. Top. D. N.I., pp. 1–28.

PHILIPPI, H. – 1932 – De Westkust van Sumatra ten Zuiden van Kroei (Benkoelen). De Mijningenieur, XIV, pp. 26–27.

PONTOPPIDAN, H. – 1913 – Verslag van een reis naar het eiland Enggano. Jaarb. Mijnwezen N.I., Verh., pp. 36–38.

RAALTEN, C. H. van – 1937 – Geologische kaart van Sumatra, schaal 1 : 200,000. Toelichting bij blad 7 (Bintoehan). Dienst v. d. Mijnbouw N.I.

RIELE, N. J. te – 1937 – Verslag bodemkundige kaartering Loeboek Linggau-gebied. Unpubl. Report Bodemk. Inst. Bogor.

RITSEMA, A. R. – 1952 – Over diepe aardbevingen in de Indische Archipel. Thesis, Utrecht.

RUTTEN, L. – 1927 – Voordrachten over de geologie van Ned. Oost Indië.

RUTTEN, M. G. – 1952 – Geosynclinal subsidence versus glacially controlled movements in Java and Sumatra. Geol. en Mijnbouw, XIV, pp. 211–220.

RUTTNER, F. – 1935 – Kieselgur and andere lakustrische Sedimente im Tobagebiet; Ein Beitrag zur Geschichte des Tobasees in Nordsumatra. Archiv f. Hydrobiol., Suppl. XIII, Tropische Binnengewässer, V, pp. 399–461.

SAPPER, K. – 1927 – Vulkankunde.

SAPPER, K. – 1935 – Geomorphologie der feuchten Tropen. Geogr. Schriften, 7.

SCHMIDT, C. – 1901 – Observations géologigues à Sumatra et à Bornéo. Bull. Soc. Géol. de France, 4e série, T.I., pp. 260–267.

SCHMIDT, F. H. and J. H. A. FERGUSON – 1952 – Rainfall types based on wet and dry period ratios for Indonesia with Western New Guinea. Verh. 42 Djaw. Meteor. Djakarta.

SCHRÖDER, E. E. W. G. – 1917 – Nias. 2 Vols.

SCHUITENVOERDER, H. J. K. – 1919 – De vulkaan Kaba. Jaarversl. Top. D. N.I., pp. 149–169.

SCHÜRMANN, H. M. E. - 1928 - Kjökkenmöddinger en paleolithicum in N. Sumatra. De Mijningenieur, 9, pp. 235-243.
SCHÜRMANN, H. M. E. - 1930 - Geologische notities uit de Bataklanden (N. Sumatra). De Mijningenieur, 11, pp. 197-200.
SEWANDONO, M. - 1938 - Het veengebied van Bengkalis. Tectona, XXXI, pp. 99-135.
SITTER, L. U. de - 1941 - Facies analyse. Geol. en Mijnbouw, 3, pp. 225-237.
SMIT SIBINGA, G. L. - 1932 - Tertiary virgations on Java and Sumatra, their relation and origin. Proc Kon. Akad. v. Wetensch. A'dam, 35.
SMIT SIBINGA, G. L. - 1933 - The Malay double (triple) orogen, I, II, III. Proc. Kon. Akad. v. Wetensch. A'dam, 36.
SMIT SIBINGA, G. L. - 1949 - Pleistocene eustasy and glacial chronology in Java and Sumatra. Verh. Kon. Ned. Geol. Mijn. Gen., Geol. serie, XV, pp. 1-31.
SMIT SIBINGA, G. L. - 1951 - On the origin and age of the peneplain of Palembang (Sumatra) Geol. en Mijnbouw, XIII, pp. 5-15.
SMIT SIBINGA, G. L. - 1952 - Interference of glacial eustasy with crustal movements and rythmic sedimentation in Java and Sumatra. Geol. en Mijnbouw, XIV, pp. 220-225.
STAUFFER, P. H. - 1966/'67 - Quaternary deposits of West Malaysia, a review. The Malayan Scientist, 3, pp. 51-57.
STEENHUIS, J. E. - 1916 - Verslag van een reis naar het eiland Enggano. Tijdschr. Kon. Ned. Aardr. Gen., XXXIII, pp. 312.
STEENIS, C. G. G. J. van - 1937 - De kalksinterterrassen van Tinggi Radja, S.O.K. Natuur in Indië, pp. 9-14.
STEENIS, C. G. G. J. van - 1938 - Exploraties in de Gajolanden. Alg. resultaten der Losir expeditie. Tijdschr. Kon. Ned. Aardr. Gen., 55, pp. 728-801.
STEHN, Ch. E. - 1934 - Die semivulkanischen Explosionen des Pematang Bata in der Soeoh-Senke (S. Sumatra) im Jahre 1933. Natuurk. Tijdschr. N.I., 94, pp. 46-69.
STEHN, Ch. E. - 1939 - De nieuwe dieptekaart van het Tobameer (N. Sumatra) van Drost & Beckering. De Ingenieur in N.I., 6, IV, pp. 121-126.
STEIGER, H. G. von - 1920 - Resultaten van geol. mijnb. verkenningen in een gedeelte van Midden Sumatra. Jaarb. Mijnwezen N.I., Verh. I, pp. 87-200.
SZEMIAN, J. - 1929/'30 - Beginselen en werkwijze der agrogeologische opname van Sumatra. Alg. Landbouw. weekbl. in N.I., 14, II, pp. 503-509.
SZEMIAN, J. - 1930 - Aanteekeningen van een agrogeologische verkenningsreis door het gebied der residentie Palembang. De Bergcultures, 4, pp. 79-81; 107-110.
SZEMIAN, J. - 1933 - Die systematische Bodenkartierung von Sumatra. Soil. Research, III, pp. 202-221.
TAVERNE, H. H. M. - 1921 - Bijdragen tot de geologie van de Gajo-lesten en aangrenzende gebieden. Jaarb. Mijnwezen N.I., Verh. I.
TERPSTRA, H. - 1932 - Voorlopige mededeeling over een geologische verkenningstocht op de eilanden Siberoet en Sipoera (Mentawei eilanden). De Mijningenieur, 13, pp. 16-20.
TERPSTRA, H. - 1936 - De richting van het materiaal transport langs de Westkust van Sumatra. De Ingenieur in N.I., IV, pp. 179-180.
TISSOT VAN PATOT, A. - 1920 - De hydrographie van het Goemai-gebergte. Jaarverslag Top. D. N.I. XV, 2, pp. 120-129.
TJIA, H. D. - 1967 - Volcanic Lineaments in the Indonesian Island Arcs. Bull. Volcanologigue, D. N.I., XV, 2, pp. 120-129.
TJIA, H. D. - 1969 - Java Sea. Encyclopaedia of Geomorphology, pp. 424-429.
TJIA, H. D. - 1970 a - Nature of displacements along the Semangko Fault zone, Sumatra. J. Trop. Geogr., 30, pp. 63-67.
TJIA, H. D. - 1970 b - Quaternary shore lines of the Sunda land, SE Asia. Geol. en Mijnbouw, 49, pp. 135-144.

TJIA, H. D. – 1970 c – Rates of diastrophic movement during the Quaternary in Indonesia. Geol. en Mijnbouw, 49, pp. 335–338.

TJIA, H. D. – 1970 d – Topographic lineaments in Riau and Lingga archipelagoes, Indonesia: their structural significance.

TJIA, H. D. et al – 1968 – Coastal accretion in western Indonesia. Bull. NIGM Bandung, 1, pp. 15–45.

TOBLER, A. – 1906 – Topographische und geologische Beschreibung der Petroleumgebiete von Moara Enim (S. Sumatra). Tijdschr. Kon. Ned. Aardr. Gen., 23, pp. 33–82.

TOBLER, A. – 1910 – Voorloopige mededeeling over de geologie der Residentie Djambi. Jaarb. Mijnwezen N.I., pp. 1–29.

TOBLER, A. – 1919 – Djambi verslag. Uitkomsten van het geol. mijnb. onderzoek in de Residentie Djambi 1906–1912, Jaarb. Mijnwezen N.I., Verh. III + atlas.

TODD, J. U. – 1939 – Palaeontological correlation between Simalur, Nias and the Banjak islands (West coast Sumatra). Unpubl. Report B.P.M.

TUYN, J. van – 1931 – Geologische kaart van Sumatra, schaal 1 : 200,000. Toelichting bij blad 4 (Soekadana). Dienst v. d. Mijnbouw N.I.

TUYN, J. van – 1932 a – De Batoemanik van Oost Palembang en de noordelijke Lampongsche Districten. De Mijningenieur, 13, pp. 20–22.

TUYN, J. van – 1932 b – Over een recente daling v. d. zeespiegel in Ned. Ind. Tijdschr. Kon. Ned. Aardr. Gen., 49, pp. 89–99.

TUYN, J. van – 1934 a – Geologische kaart van Sumatra, schaal 1 : 200,000. Toelichting bij blad 13 (Wiralaga). Dienst v. d. Mijnbouw N.I.

TUYN, J. van – 1934 b – Geologische kaart van Sumatra, schaal 1 : 200,000. Toelichting bij blad 8 (Menggala). Dienst v. d. Mijnbouw N.I.

UBAGHS, J. G. H. – 1941 a – Geoology of the Lampong districts. Unpubl. Report Dienst Mijnwezen N.I.

UBAGHS, J. G. H. – 1941 b – De geologie van Benkoelen. Unpubl. Report Dienst Mijnwezen N.I.

UMBGROVE J. H. F. – 1931 a – The Sibajak volcano. Ztschr. f. Vulkankunde, XIII, pp. 237–244.

UMBGROVE, J. H. F. – 1931 b – De koraalriffen van Emmahaven (W. Sumatra). Leidsche Geol. Meded., IV, 1, 3, pp. 9–24.

UMBGROVE, J. H. F. – 1932 – Het neogeen in den Indischen Archipel. Tijdschr. Kon Ned. Aardr. Gen., 49, pp. 769–834.

UMBGROVE, J. H. F. – 1938 – Geological History of the East Indies. Bull. Am. Ass. of Petrol. Geol., 22, pp. 1–70.

UMBGROVE, J. H. F. – 1949 – Structural history of the East Indies.

VALKENBURG, S. van – 1921 – Geomorphologische beschouwingen over de Padangsche bovenlanden. Jaarverslag Top. D. N.I., pp. 1–30.

VEEN, A. L. W. E. van – 1913 – Bijdrage tot de geologie van Nias. Samml. Reichsm. Geol. Leiden, 9, pp. 225–243.

VENING MEINESZ, F. A. – 1954 – Indonesian archipelago: a geophysical study. Bull. Geol. Soc. Amer., 65, pp. 143–164.

VERBEEK, R. D. M. – 1876 – Geologische beschrijving van het eiland Nias. Jaarb. Mijnwezen N.I., I, pp. 3–13.

VERBEEK, R. D. M. – 1883 – Topografische en geologische beschrijving van een gedeelte van Sumatra's Westkust. Jaarb. Mijnwezen N.I.

VERBEEK, R. D. M. – 1897 – Geologische beschrijving van Bangka en Billiton. Jaarb. Mijnwezen N.I. (+ atlas).

VERSTAPPEN, H. Th. – 1955 a – Geomorphic notes on Kerintji (Central Sumatra). Indon. J. f. Nat. Sci., III, pp. 166–177.

VERSTAPPEN, H. Th. – 1955 b – Geomorphologische Notizen aus Indonesiën. Erdkunde, IX, pp. 134–144.
VERSTAPPEN, H. Th. – 1956 – The physiographic basis of pioneer settlement in southern Sumatra. Publ. 6, Balai Geografi, Djatop. Djakarta.
VERSTAPPEN, H. Th. – 1960 – Some observations on karst development in the Malay archipelago. J. of Trop. Geogr., XIV, pp. 1–10.
VERSTAPPEN, H. Th. – 1961 – Some 'volcano-tectonic' depressions of Sumatra, their origin and mode of development. Proc. Kon. Akad. v. Wetensch. A'dam, B, 64 pp. 428–443.
VERSTAPPEN, H. Th. – 1964 a – Geomorphology of Sumatra, J. of Trop. Geogr., XVIII, pp. 184–191.
VERSTAPPEN, H. Th. – 1964 b – Geomorphology in Delta Studies, ITC-publ. B. 24.
VETH, P. J. – 1881/'84 – Midden-Sumatra. Reizen en onderzoekingen der Sumatra-expeditie uitgerust door het Aardrijksk. Genootschap 1877–1879.
VOLZ, W. – 1899 – Beiträge zur geologischen Kenntniss von Nord-Sumatra. Thesis Breslau.
VOLZ, W. – 1909 – Nord Sumatra. 2 Vols.
WAARD, D. de and Th. H. F. KLOMPE – 1952 – The recent activity of G. Marapi in Central Sumatra. Indon. J. f. Nat. Sci., I, pp. 131–140.
WASTL, L. – 1939 – Prähistorische Menschenreste aus dem Muschelhügel von Bindjai-Tamiang in N. Sumatra; Festschrift O. Reches.
WESTERVELD, J. – 1931 – Geologische kaart van Sumatra, schaal 1 : 200,000. Toelichting bij blad 5 (Kotaboemi). Dienst v. d. Mijnbouw N.I.
WESTERVELD, J. – 1933 – Geologische kaart van Sumatra, schaal 1 : 200,000. Toelichting bij blad 3 (Bengkoenat). Dienst v. d. Mijnbouw N.I.
WESTERVELD, J. – 1936 – On the geology of N. Banka (Djeboes). Proc. Kon. Akad. v. Wetenschap A'dam., 39, pp. 1122–1131.
WESTERVELD, J. – 1941 a – De tektonische bouw van Z. Sumatra. Hand. 28e Nat. Geneesk. Congr. Utrecht, 4e afd., pp. 264–267.
WESTERVELD, J. – 1941 b – Three geological sections across S-Sumatra. Proc. Kon. Akad. v. Wetensch. A'dam, 44, pp. 1131–1139.
WESTERVELD, J. – 1943 – Welded rhyolitic tuffs or 'ignimbrites' in the Pasoemah region, S. Sumatra. Leidsche Geol. Meded., 13, pp. 202–217.
WESTERVELD, J. – 1947 – On the origin of the acid volcanic rocks around Lake Toba, N. Sumatra. Verh. Kon. Akad. v. Wetensch. A'dam, Afd. Natuurk., II, 43, pp. 1–51.
WESTERVELD, J. – 1952 – Quaternary Volcanism on Sumatra. Bull. Geol. Soc. Amer., 63, pp. 561–594.
WESTERVELD, J. – 1953 – Eruptions of acid pumice tuffs and related phenomena along the Sumatran fault trough system. Proc. 7th Pac. Sci. Congr. II, pp. 411–438.
WICHMANN, A. – 1904 – Über die Vulkane von Nord-Sumatra. Ztschr. d. D. Geol. Ges. 56, pp. 230–232.
WING EASTON, N. – 1894 – Een geologische verkenning in de Toba landen. Jaarb. Mijnwezen N.I., Wetensch. Ged., pp. 99–164.
WING EASTON, N. – 1896 – Der Tobasee. Ztschr. d. D. Geol. Ges., 48, pp. 435–467.
WITKAMP, H. – 1922 – Het profiel van den Piek van Kerintji. Natuurk. Tijdschr. N.I., 82, pp. 215–221.
ZEN, M. T. – 1969 – The state of Anak Krakatau in September 1968. Bull. NIGM Bandung, 2, pp. 15–23.
ZWIERZYCKI, J. – 1916 – Geologische beschrijving van het eiland Poeloe We. Jaarb. Mijnwezen N.I., Verh. 2, pp. 5–6.
ZWIERZYCKI, J. – 1919 – Verslag van een reis door Sumatra. Unpubl. Report Geol. Mijnb. D., Bandoeng.
ZWIERZYCKI, J. – 1920 a – Het jong-tertiaire gebied van NW-Atjeh (Atjeh III), Jaarb. Mijnwezen, Verh. I.

ZWIERZYCKI, J. – 1920 b – Zijn de indische petroleumterreinen, in het bijzonder die op Sumatra peneplains of abrasievlakken? De Mijningenieur, I, pp. 3–5.

ZWIERZYCKI, J. – 1922/'30 – Geologische overzichtskaart van den Ned. Oost Indischen Archipel ! : 1,000,000. Toelichting bij de bladen I, VII, VIII.

ZWIERZYCKI, J. – 1931 – Geologische kaart van Sumatra, schaal 1 : 200,000. Toelichting bij blad 1 (Teloekbetoeng). Dienst v. d. Mijnbouw N.I.

ZWIERZYCKI, J. – 1932 – Geologische kaart van Sumatra, schaal 1 : 200,000. Toelichting bij blad 2 (Kotaägoeng). Dienst v. d. Mijnbouw N.I.